Laser as a Detection: From Spectral Imaging to LiDAR for Remote Sensing Applications

Laser as a Detection: From Spectral Imaging to LiDAR for Remote Sensing Applications

Guest Editors

Jianfeng Chen
Ming Zhao
He Tian

MDPI

Basel • Beijing • Wuhan • Barcelona • Belgrade • Novi Sad • Cluj • Manchester

Guest Editors
Jianfeng Chen
State Key Laboratory of Laser Interaction with Matter
Chinese Academy of Sciences
Hefei
China

Ming Zhao
School of Electronic Engineering
Huainan Normal University
Huainan
China

He Tian
College of Science
Northeast Forestry University
Harbin
China

Editorial Office
MDPI AG
Grosspeteranlage 5
4052 Basel, Switzerland

This is a reprint of the Special Issue, published open access by the journal *Photonics* (ISSN 2304-6732), freely accessible at: https://www.mdpi.com/journal/photonics/special_issues/1CT0JYUA49.

For citation purposes, cite each article independently as indicated on the article page online and as indicated below:

Lastname, A.A.; Lastname, B.B. Article Title. *Journal Name* **Year**, *Volume Number*, Page Range.

ISBN 978-3-7258-5283-3 (Hbk)
ISBN 978-3-7258-5284-0 (PDF)
https://doi.org/10.3390/books978-3-7258-5284-0

Contents

About the Editors

Jianfeng Chen

Jianfeng Chen works at the State Key Laboratory of Laser Interaction with Matter, Anhui Institute of Optics and Fine Mechanics, Hefei Institutes of Physical Science, Chinese Academy of Sciences, Hefei, China. His work focuses on atmospheric Lidar design and optical engineering for wind and trace-gas observations, with emphasis on calibration, stability control, and scanning/overlap optimization. He has led projects on Rayleigh-Doppler wind Lidar and differential-absorption techniques, conducted field validations, and published widely in remote sensing and photonics journals.

Ming Zhao

Ming Zhao works at the School of Electronic Engineering, Huainan Normal University, Huainan, China. His research spans remote sensing, Lidar, laser spectroscopy, and atmospheric environment studies. He develops Lidar and spectroscopic instruments and retrieval algorithms, integrates multi-source observations for environmental assessment, and collaborates on application-driven field campaigns. His publications cover instrument design, signal processing, and atmospheric measurement methods.

He Tian

He Tian works at the College of Science, Northeast Forestry University, Harbin, China. His research centers on fiber-optic sensing, optical waveguide devices, and optoelectronic spectral applications. He designs and models integrated photonic structures—including plasmonic and waveguide sensors—for refractive-index and multi-parameter detection, and explores their translation to biochemical and environmental monitoring. He has authored peer-reviewed articles on device physics and sensing.

Preface

This Reprint surveys state-of-the-art research on laser-based sensing for remote sensing, with emphasis on spectral imaging and Lidar. Its scope spans the ultraviolet to the mid-infrared and covers instrumentation; system design; retrieval algorithms; calibration and standardization; deployment strategies; and data processing. The aim is to provide a coherent view of how advances in photonics and computation are transforming measurements of the atmosphere, land, ocean, and urban environments.

Our motivation is twofold. First, society increasingly depends on precise, traceable observations to support air-quality management, climate and weather services, natural-hazard response, resource assessment, and autonomous sensing. Second, rapid progress in compact lasers, integrated photonics, and real-time analytics enables rugged, networked systems to transition from laboratory prototypes to sustained field operation. The papers collected here document this transition through laboratory validation, field campaigns, and comparative studies, and they highlight themes of miniaturization, stability, calibration, multimodal data fusion, and interoperability.

This Reprint addresses researchers and engineers in photonics, remote sensing, environmental science, and robotics, as well as graduate students, industrial practitioners, and public agencies seeking reliable optical tools for environmental and industrial monitoring. By bringing together fundamentals, instrument architectures, retrieval methods, and application case studies, this Reprint aims to serve as a practical reference and a catalyst for cross-disciplinary collaboration.

Jianfeng Chen, Ming Zhao, and He Tian
Guest Editors

photonics MDPI

Editorial

Editorial: Frontiers and Applications of Laser Detection—From Spectral Imaging to Lidar Remote Sensing

Jianfeng Chen [1,*], Ming Zhao [2] and He Tian [3]

[1] State Key Laboratory of Laser Interaction with Matter, Anhui Institute of Optics and Fine Mechanics, Hefei Institutes of Physical Science, Chinese Academy of Sciences, Hefei 230031, China
[2] School of Electronic Engineering, Huainan Normal University, Huainan 232063, China; zhaom@hnnu.edu.cn
[3] College of Science, Northeast Forestry University, Harbin 150040, China; tianhe@nefu.edu.cn
* Correspondence: jfchen@aiofm.ac.cn

The rapid advancement of optoelectronics and precision measurement technologies has made laser detection an essential and pioneering tool in contemporary remote sensing. Ranging from highly sensitive spectral imaging and Lidar with exceptional spatial resolution to innovative applications in environmental monitoring and biomedical imaging, laser detection has fundamentally transformed the methods by which we acquire, interpret, and utilize information about our environment [1–3]. This Special Issue, titled "Laser Detection: Remote Sensing Applications from Spectral Imaging to Lidar," presents recent technological breakthroughs and a diverse range of application achievements in the field, highlighting the broad potential of laser detection technologies in both basic research and engineering applications.

In recent years, laser detection technology has achieved substantial advancements in wavelength coverage, sensitivity, system miniaturization, and intelligent data processing [4–7]. The advent of next-generation spectral lasers, broadband detectors, and integrated optical components has endowed laser spectroscopic imaging with distinct advantages in molecular identification and quantitative analysis [8–10]. Simultaneously, Lidar has become a versatile tool, widely used in atmospheric remote sensing, surface mapping, ecological monitoring, autonomous driving, and space exploration [11–13]. Despite these advances, several major challenges remain in this field. Key issues include further improving the spatial and temporal resolution of spectroscopic imaging, suppressing background interference, and achieving efficient integration and cross-calibration of data from multiple platforms and sensors [14,15]. Furthermore, addressing weak signal detection and multi-parameter retrieval in complex environments will require the synergistic optimization of algorithms and system architectures—a key direction for future research [16,17].

The articles in this Special Issue encompass a broad spectrum of topics, including novel mechanisms for laser-based atmospheric transport and absorption detection, innovations in Lidar technology and signal processing, advances in detector arrays and system calibration, developments in multi-source and multi-modal lasers and imaging, applications in atmospheric remote sensing and environmental monitoring, and emerging trends and engineering applications in the field. Some studies focus on reconstructing atmospheric transport and absorption fields, proposing new laser transmission models and gas absorption imaging techniques that offer theoretical and technical foundations for high-precision environmental monitoring and flow field diagnostics. Several papers address system optimization and algorithmic innovation for Lidar in atmospheric profiling, wind field measurement, and weak target detection, thereby advancing the core applications of Lidar in environmental and climate remote sensing. In parallel, advancements in detector array consistency calibration, low-cost imaging systems, underwater suppression and scattering lasers, and cross-disciplinary innovations in novel optical materials for nano-sensing have broadened the application boundaries of laser systems. This issue further highlights cutting-edge advances in AI-based noise reduction for laser signals, multi-modal fusion and system intelligence, and system miniaturization, reflecting the deep integration of laser detection theory and engineering practice. Collectively, these achievements have enriched and advanced the field of laser detection, providing a robust foundation for future multidisciplinary collaboration and practical innovation.

Looking ahead, laser detection is expected to achieve further breakthroughs in areas such as multi-scale integration and intelligent processing, the development of high-throughput portable instruments, cross-disciplinary collaboration and standardization, and the exploration of emerging application domains [18–20]. Specifically, the integration of multi-band and multi-modal laser signals with artificial intelligence algorithms will further enhance detection sensitivity and quantitative capabilities in complex environments. The continued implementation of integrated, miniaturized, and cost-effective laser detection systems will expand their applications to emerging scenarios such as mobile observation and emergency monitoring [21]. The convergence of disciplines—including optics, electronics, computing, and environmental science—will accelerate data standardization and platform interoperability, facilitating the widespread deployment and international collaboration of laser detection technologies [22,23]. Moreover, laser detection will progressively expand into frontier areas such as ecological monitoring, biomedicine, and urban intelligent sensing, thereby providing robust support for the development of a digital Earth and a smart society [24,25].

This Special Issue presents the latest advances in laser detection, spanning from fundamental principles to engineering applications, and highlights the deep integration and innovation-driven development of this technological domain. We anticipate that these studies will further stimulate interest and investment from both the academic community and industry, thereby advancing the value of laser detection technology in scientific discovery and societal applications.

Acknowledgments: The Guest Editors of this Special Issue, "Laser as a Detection: From Spectral Imaging to LiDAR for Remote Sensing Applications", would like to express our sincere thanks and deep appreciation to all authors published in this Special Issue for their contribution to its success. We also thank our reviewers, as well as the Photonics editors and staff for their outstanding support. During the preparation of this manuscript the author used OpenAI for the purposes of language polishing and literature search. The authors have reviewed and edited the output and take fully responsible for the content of this publication.

Conflicts of Interest: The authors declare no conflicts of interest.

References

1. Xu, F.; Qiao, D.; Xia, C.; Song, X.; Zheng, W.; He, Y.; Fan, Q. A semi-coaxial MEMS LiDAR design with independently adjustable detection range and angular resolution. *Sens. Actuators A Phys.* **2021**, *326*, 112715. [CrossRef]
2. Zhang, H.-S.; Qiao, L.-L.; Cheng, Y. Air lasing: Remote high-resolution spectroscopy for atmospheric sensing. *Acta Phys. Sin.* **2022**, *71*, 233401. [CrossRef]
3. Ran, Y.; Song, S.; Hou, X.; Chen, Y.; Chen, Z.; Gong, W. Multi-echo hyperspectral reflectance extraction method based on full waveform hyperspectral LiDAR. *ISPRS J. Photogramm. Remote Sens.* **2024**, *207*, 43–56. [CrossRef]
4. Li, A.-W.; Shan, T.-Q.; Guo, Q.; Pan, X.-P.; Liu, S.-R.; Chen, C.; Yu, Y.-S. Fiber Fabry-Perot interferometric high-temperature sensors for aerospace monitoring. *Chin. Opt.* **2022**, *15*, 609–624. [CrossRef]
5. Han, M.-M.; Wei, H.-Y.; Zou, W.; Meng, L.; Zhang, M.-P.; Meng, X.; Chen, W.-W.; Shao, H.; Wang, C.-J. Rapid on-site detection of coumatetralyl in environmental water based on surface-enhanced Raman spectroscopy. *Environ. Chem.* **2023**, *42*, 1524–1532. [CrossRef]

6. Liu, W.-Q. Opportunities and Challenges for Development of Atmospheric Environmental Optics Monitoring Technique Under "Double Carbon" Goal. *Acta Opt. Sin.* **2022**, *42*, 0600001. [CrossRef]
7. Yang, S.; Kim, J.; Swartz, M.E.; Eberhart, J.K.; Chowdhury, S. DMD and microlens array as a switchable module for illumination angle scanning in optical diffraction tomography. Biomed. *Opt. Express* **2024**, *15*, 5932–5946. [CrossRef]
8. Zhang, F.; Xie, H.; Yuan, L.; Zhang, Z.; Fu, B.; Yu, S.; Li, G.; Zhang, N.; Lu, X.; Yao, J.; et al. Background-free single-beam coherent raman spectroscopy assisted by air lasing. *Optics Letter.* **2022**, *47*, 481–484. [CrossRef] [PubMed]
9. Wang, L.; Zhang, Y.; Liu, J.; Chen, B.; Zhou, M. Mid-infrared dual-comb spectroscopy for multi-component gas analysis in industrial emissions. *Appl. Sci.* **2021**, *11*, 3660. [CrossRef]
10. Tang, D.; Li, Z.; Xia, H. Clustering of weak fluorescence spectra from bioaerosol in air using laser-induced fluorescence lidar. *Opt. Express* **2025**, *33*, 24396–24412. [CrossRef]
11. Wen, H.-Y.; Weng, Y.-Q.; Chen, R.-Y.; Hsu, H.-C.; Yeh, Y.-T.; Chiang, C.-C. A double helix-shaped optical fiber sensor for non-endoscopic diagnosis of gastrin-17. *Analyst* **2022**, *147*, 4562–4569. [CrossRef]
12. Zhou, J.; Xia, Y.-P.; Li, H.-H.; Wang, L.; Zhang, K.; Sun, Q. OrchardQuant-3D: Drone-LiDAR fusion for 3D phenotyping of floral traits. *Plant Biotechnol. J.* **2025**, *23*, 1254–1270. [CrossRef]
13. Haierxin Photonics R&D Team; Liu, X.; Zhang, F.; Li, Q. Mid-infrared quantum cascade lasers for non-invasive medical diagnostics. *J. Biophotonics* **2025**, *18*, e202400123. [CrossRef]
14. Chen, P.-P.; Whitfield, C.; Zhang, F.-H.; Li, X.; Wang, G. LPVIMO-SAM: Multi-sensor SLAM with polarization vision for degraded environments. *IEEE Robot. Autom. Lett.* **2025**, *10*, 4500–4507.
15. Shi, B.; Zheng, M.-Y.; Hu, Y.; Zhao, Y.; Shang, Z.; Zhong, Z.; Chen, Z.; Luo, Y.-H.; Long, J.; Sun, W.; et al. A hyperfine-transition-referenced vector spectrum analyzer for visible-light integrated photonics. *Nat. Commun.* **2025**, *16*, 61970. [CrossRef] [PubMed]
16. Beer, M.; Haase, J.; Charbon, E. Background Light Rejection in SPAD-Based LiDAR Sensors by Adaptive Photon Coincidence Detection. *Sensors* **2018**, *18*, 4338. [CrossRef]
17. Carreón, R.V.; Rodríguez-Hernández, A.G.; Serrano de la Rosa, L.E.; Gervacio-Arciniega, J.J.; Krishnan, S.K. Mechanically Flexible, Large-Area Fabrication of Three-Dimensional Dendritic Au Films for Reproducible Surface-Enhanced Raman Scattering Detection of Nanoplastics. *ACS Sens.* **2025**, *10*, 1747–1755. [CrossRef]
18. Svanberg, S.; Kaldvee, J.; Andersson, M.; Persson, L. Laser spectroscopy in ecological and medical applications: From leaf photosynthesis to cancer diagnostics. *J. Biomed. Opt.* **2021**, *26*, 080601. [CrossRef]
19. Chen, M.; Liu, D.; Qiao, L.; Zhou, P.; Feng, J.; Ng, K.W.; Liu, Q.; Wang, S.; Pan, H. In-situ/operando Raman techniques for in-depth understanding of electrocatalysis. *Chem. Eng. J.* **2023**, *457*, 141280. [CrossRef]
20. Sigernes, F.; Dyrland, M.; Peters, N.; Lorentzen, D.; Baddeley, L. Hyperspectral imaging from unmanned aerial vehicles for Arctic environmental monitoring. *Remote Sens. Environ.* **2023**, *287*, 113482. [CrossRef]
21. Zhang, J.; Wang, C.; Chen, Y.; Xiang, Y.; Huang, T.; Shum, P.P.; Wu, Z. Fiber structures and material science in optical fiber magnetic field sensors. *Front. Optoelectron.* **2022**, *15*, 34. [CrossRef]
22. Chen, B.; Xu, T.; Wang, S.; Li, Y.; Zhang, G. Optical fiber sensors: Principles and applications in smart infrastructure. *Sensors* **2022**, *22*, 8793. [CrossRef]
23. Kar, J.; Vaughan, M.A.; Lee, K.-P.; Tackett, J.L.; Avery, M.A.; Garnier, A.; Getzewich, B.J.; Hunt, W.H.; Josset, D.; Liu, Z.; et al. CALIPSO lidar calibration at 532 nm: Version 4 nighttime algorithm. *Atmos. Meas. Tech.* **2018**, *11*, 1459–1479. [CrossRef] [PubMed]
24. Zhang, F.; Huang, X.; Ren, X.; Sun, Y.; Liu, Z. Underwater lidar based on blue-green supercontinuum laser for marine particle profiling. *Opt. Lasers Eng.* **2025**, *166*, 107612. [CrossRef]
25. Brosseau, C.L.; Colina, A.; Perales-Rondon, J.V.; McCabe, E.M.; Smith, W.E.; Ren, B.; Wang, X. Electrochemical surface-enhanced Raman spectroscopy. *Nat. Rev. Methods Primers* **2023**, *3*, 79. [CrossRef]

photonics

MDPI

Review

Evolution of Single Photon Lidar: From Satellite Laser Ranging to Airborne Experiments to ICESat-2

John J. Degnan †

Independent Researcher, Odenton, MD 21113, USA; jjdegnan3rd@gmail.com

† The author is a semi-retired technical consultant who previously held several research and supervisory positions at NASA Goddard Space Flight Center (1964–2003) and served as Chief Scientist at Sigma Space Corporation (2003–2019) before entering semi-retirement as a part-time consultant. He also taught a two-semester graduate course in Quantum Electronics as a Distinguished Adjunct Professor of Physics at the American University in Washington, DC, from 1989 to 1993. He co-founded the International Laser Ranging Service (ILRS) in 1997 and served as its first Governing Board Chairman from 1998 to 2002. He is a Fellow of Optica (formerly the Optical Society of America) and the International Association for Geodesy, an IEEE Senior Life Member, and a member of the Sigma Pi Sigma Physics and the Drexel 100 Honor Societies.

Abstract: In September 2018, NASA launched the ICESat-2 satellite into a 500 km high Earth orbit. It carried a truly unique lidar system, i.e., the Advanced Topographic Laser Altimeter System or ATLAS. The ATLAS lidar is capable of detecting single photons reflected from a wide variety of terrain (land, ice, tree leaves, and underlying terrain) and even performing bathymetric measurements due to its green wavelength. The system uses a single 5-watt, Q-switched laser producing a 10 kHz train of sub-nanosecond pulses, each containing 500 microjoules of energy. The beam is then split into three "strong" and three "weak" beamlets, with the "strong" beamlets containing four times the power of the "weak" beamlets in order to satisfy a wide range of Earth science goals. Thus, ATLAS is capable of making up to 60,000 surface measurements per second compared to the 40 measurements per second made by its predecessor multiphoton instrument, the Geoscience Laser Altimeter System (GLAS) on ICESat-1, which was terminated after several years of operation in 2009. Low deadtime timing electronics are combined with highly effective noise filtering algorithms to extract the spatially correlated surface photons from the solar and/or electronic background noise. The present paper describes how the ATLAS system evolved from a series of unique and seemingly unconnected personal experiences of the author in the fields of satellite laser ranging, optical antennas and space communications, Q-switched laser theory, and airborne single photon lidars.

Keywords: single photon lidar; satellite laser ranging; Q-switched microlasers; array detectors; noise filters; 3D imaging; scanners; ICESat-2

1. Introduction

Theodore Maiman at the Hughes Research Laboratory in California reported the first ruby laser in 1960 [1], and the National Aeronautics and Space Administration (NASA) in the USA was quick to identify and implement potential space applications. In 1964, Dr. Henry H. Plotkin, Head of the Instrument Electro-optics Branch at the NASA Goddard Space Flight Center (GSFC), led a team that, on 31 October 1964, measured the distance to a satellite, Explorer 22B, by transmitting Q-switched ruby laser pulses at 1 Hz to a collection of retroreflectors mounted on the satellite, which reflected the pulse back to the ground station [2]. The telescope, located at the Goddard Optical Research Facility (GORF) a few miles from GSFC, was guided by two individuals seated on the telescope and independently controlling the azimuth and elevation axes to keep the telescope crosshairs on the sunlit satellite at night. By measuring the roundtrip time of flight (TOF) and multiplying by the speed of light, one could then compute the distance to the satellite for precise orbit determination. At the time, I was a first-year cooperative work/study student at GSFC and was honored to serve as a junior member of Dr. Plotkin's SLR team following my freshman year at Drexel Institute of Technology (now Drexel University) in Philadelphia. Upon graduating from Drexel with a B.S. in Physics in 1968, I was offered a permanent position in Dr. Plotkin's branch. Daylight ranging using computer-driven telescopes was first achieved by GSFC's Don Premo in 1969. The earliest Satellite Laser Ranging (SLR) measurements were accurate to about 2 m, representing roughly a factor of 40 improvement relative to earlier radar measurements. The accuracy was limited by a number of factors, including the temporal width of the Q-switched ruby laser pulse (~1 μs), variable propagation times within the early photomultiplier tube detectors which amplified the weak return signals, the precision of the electronics measuring the time interval between the start and stop event, the temporal broadening of the return pulse caused by the spatial distribution of the retroreflectors on the satellite, and uncertainties in delays caused by the intervening atmosphere.

In 1969, Professor Carroll O. Alley, a physics professor at the University of Maryland in College Park, led a joint NASA/university team of researchers in the first successful attempt to measure the distance to an array of reflectors landed on the Moon by NASA's Apollo 11 astronauts [3]. Over the next three decades that followed, SLR ranging errors were reduced to a few mm [4] through the use of the following:

- frequency-doubled mode-locked Nd:YAG laser transmitters or Q-switched microlasers with sub-nanosecond pulse widths at a wavelength of 532 nm;
- MicroChannel Plate PhotoMultipliers (MCP/PMTs), which, unlike earlier PMTs, could record and amplify single photons while tightly restricting the electron path within the detector;
- highly accurate Time Interval Units (TIUs) or Event Timers (ETs) such as the HP5370 produced by Hewlett Packard;
- a calibration target mounted on a post near the SLR station with the distance between the target and the telescope "origin" (defined as the intersection of the telescope azimuth and elevation axes) periodically surveyed with mm accuracy and monitored for consistency over time;
- and finally, collocated meteorological instruments that monitored the atmospheric pressure and temperature at the SLR station to help account for changing atmospheric delays.

Satellite retroreflector array designs have also evolved over time in order to enhance the received signal strength while simultaneously minimizing the temporal spreading of the reflected pulse and/or the sensitivity of the measurement to the array angle of attack. This can be accomplished by distributing retroreflectors over (1) the surface of passive spherical satellites (e.g., Starlette, LAGEOS, etc.) used in the study of the Earth's gravity field, tectonic plate motion, etc. or (2) retroreflectors distributed over a segment of a sphere (instead of earlier flat panels) for higher altitude observational or navigation satellites, which always have a flat surface oriented toward Earth [5].

In the late 1960s, the USA and France implemented and deployed a total of five SLR stations. As the scientific demands for SLR technology grew, additional nations sponsored a number of fixed stations, often associated with preexisting university astronomical observatories hosting meter-class telescopes. In addition, a number of mobile stations were designed to support studies of tectonic plate motion at a variety of sites, initially confined largely to North America and Europe. NASA developed a total of eight Mobile Laser (MOBLAS) stations, each of which was housed in two rather large trailers. NASA later funded the development of smaller Transportable Laser Ranging Systems (TLRS 1 through 4), which were each contained within a single, smaller trailer in order to simplify transport to remote sites such as French Polynesia and Easter Island and to move between multiple sites in Europe and North America to better study global tectonic plate motion.

NASA funded the University of Texas at Austin to develop TLRS-1, while electrical engineer Thomas S. Johnson at GSFC, a key member of Henry Plotkin's original SLR team, led the development of TLRS-2. Ultimately, TLRS-2 alternated between sites in Chile, French Polynesia, and Easter Island. Easter Island was the only above-water location on the Nazca tectonic plate and was roughly 1900 miles from the nearest land location. During the 1980s, my Advanced Electro-Optical Instrument Section was tasked by my then NASA branch head, Dr. John H. McElroy, to upgrade the MOBLAS and TLRS-2 station performance. Over time, additional transportable SLR stations were developed by France (FTLRS) [6], Germany (miniSLR), China (TROS), Korea (ARGO-M), etc.

In 1997, in order to better coordinate international SLR research activities, the author co-founded the International Laser Ranging Service (ILRS) with Dr. Bob Schutz of the University of Texas at Austin [7]. In 1998, I was selected as one of two NASA representatives to the International Governing Board (GB) and elected by the GB as its first chairman from 1998 until my retirement from NASA in 2002. Dr. Michael Pearlman of the Smithsonian Institution was elected head of the Central Bureau and still holds that position today. Presently, there are approximately 40 permanently located SLR stations, most of which are in the Northern Hemisphere. For a comprehensive review of recent SLR scientific and technological trends, the reader is referred to several comprehensive review articles published in 2019 [7–9].

2. NASA SLR2000: The First Single-Photon-Sensitive SLR Station

In 1989, I left the GSFC Engineering Directorate for the Science Directorate to serve as Deputy Manager of NASA's Crustal Dynamics Project (CDP) under Project Manager John M. Bosworth. Within a few years, I was appointed head of the newly formed Space Geodesy and Altimetry Projects Office within the same directorate. In 1994, Dr. David E. Smith, head of the Laboratory for Terrestrial Physics and an early and prolific scientific user of SLR data, requested a meeting with me. The purpose of the meeting was to discuss the possibility of reducing the cost of SLR operations in order to meet the growing science demands within the available budget. These demands included measuring the Earth's gravity field via precise satellite orbits, global tectonic plate motion, orbital support of spaceborne microwave radars and laser altimeters, as well as global navigation satellites such as GPS, GLONASS, Galileo, Beidou, etc. The major cost drivers of the existing SLR stations included (1) the use of powerful and technically complex sub-nanosecond pulse lasers, precision timing equipment, expensive quasi-meter-class, high-precision tracking telescopes and their protective domes, the need for supporting meteorological equipment for atmospheric delay correction, aircraft radars to protect airborne personnel from potentially eye-damaging beam intensities and (2) the resulting manpower costs per shift, typically three to four people, to operate and/or monitor a wide range of complex hardware. Clearly, any new cost-saving SLR station design had to (1) be highly automated, (2) utilize smaller and less expensive lasers and tracking telescopes, and (3) be able to operate in an eye-safe mode. With regard to the system automation requirement, I relied heavily on my longtime NASA colleague, Ms. Jan McGarry, a talented mathematician and computer programmer who had been assigned to my Advanced Electro-optical Instrument Section in 1979 and who thankfully followed me in the early 1990s when I transferred from GSFC Engineering to the Science Directorate.

For decades, the laser beam in SLR systems was transmitted to the target via a narrow tube running parallel to the optical axis of the receiver telescope. Any laser operating in the lowest order and least divergent spatial mode, TEM_{00}, has a Gaussian beam shape. In the far field, the TEM_{00} mode retains its Gaussian shape but has a beam diameter between $1/e^2$ intensity points given by $D(R) = 2\lambda R/\pi\omega$ where λ is the laser wavelength, R is the range to the target, and ω is the Gaussian beam waist radius where the phase front is planar, usually located within the laser itself. Clearly, as the beam diameter at the target grows with respect to the target array, fewer photons are reflected back to the receiver telescope for a given pulse energy. As a result, more powerful lasers and/or larger meter-class receive telescopes had routinely been used to obtain adequate laser returns from the most distant satellites and especially from reflectors on the Moon. This clearly resulted in more expensive systems and/or greatly increased the threat to eye safety.

The problems associated with both system costs and eye safety reminded me of research on transmitter and receiver optical antenna gain that I had supervised in the mid-1970s with a young electrical engineer, Bernard J. Klein [10,11]. The latter effort was in support of proposed spaceborne CO_2 heterodyne laser communications systems, which were designed to operate at an even more divergent infrared wavelength of 10.6 μm.

The solution was to use the existing large receive telescope diameter to transmit a much larger and therefore less divergent beam that would deposit orders of magnitude more energy/power onto either a passive target or a remote communications terminal telescope. The same holds true for large meter-class astronomical telescopes, which, to reduce the overall length, often use secondary mirrors that partially block both the transmitted and received light. As a result, the optimum expanded Gaussian beam diameter that exits the telescope and maximizes the energy incident on the satellite is a function of both the primary and secondary mirror diameters [10]. In the process, however, the radial truncation of the transmitted beam by the finite aperture of the telescope's primary mirror/lens and/or central obscuration by a secondary mirror (if any) causes the Gaussian shape to evolve into a strong central lobe surrounded by a series of increasingly weaker rings in the far field of the telescope [10]. Furthermore, the presence of a secondary mirror requires a larger Gaussian beam diameter at the primary to increase the fraction of light exiting the telescope and therefore maximize the intensity of the central lobe in the far field. At the same time, however, a secondary mirror causes a greater fraction of the transmitted laser energy to be transferred from the central lobe to the surrounding rings relative to the no secondary mirror case. In all cases, however, the intensity of the first and strongest ring is more than two orders of magnitude less than that of the central lobe. The resulting far-field gain of the optimized transmitting telescope when pointed correctly at the target is given by the following simple equation [10]:

$$G_T = \frac{4\pi A_P}{\lambda^2} g_T = \frac{4\pi A_P}{\lambda^2}\left(1.12 - 1.30\gamma^2 + 2.12\gamma^4\right) \quad (1)$$

where $A_p = \pi a^2$ is the area of the telescope primary mirror or lens, $\gamma = b/a < 0.4$ is the ratio of the secondary mirror radius b to the primary mirror radius a, and λ is the laser wavelength. The light returning from the target is also reduced by an additional factor of $g_R = (1 - \gamma^2)$ due to blockage of incoming light by the secondary mirror. The secondary mirror can also influence the light distribution produced in the detector plane [11]. The combined effects are examined in [5] for the case of a Lunar Laser Ranging (LLR) station ranging to a single retroreflector with diameter D_{cc}, and it yields the following equation for the number of photons received by the ground station:

$$n_s = n_t(\eta_t\eta_r\eta_d)T_{atm}^2\rho g_T\, g_R\left(\frac{\pi}{2}\right)^4\left(\frac{D_{cc}\ D_p}{\lambda R}\right)^4 \leq 0.034 n_t T_{atm}^2\left(\frac{D_{cc}\ D_p}{\lambda R}\right)^4 \quad (2)$$

where n_t is the number of photons transmitted per laser pulse, $\eta_t = \eta_r = 0.66$ are estimated optical efficiencies of the transmit and receive paths in the telescope, $\eta_d = 0.70$ is the assumed detector efficiency, $T_{atm}{}^2$ is the two-way transmission through the atmosphere (if any), $g_T = 1.12$ and $g_R = 1$ are the optimized geometric telescope gains in the absence of a secondary mirror, ρ is the reflectivity of the retroreflector surfaces, D_{cc} is the cube corner diameter, λ is the laser wavelength, and R is the range to the target retroreflector. The final inequality in Equation (2) assumes estimated values for the various efficiencies [5].

In summary, a much lower pulse energy, transmitted optimally through a common transmit/receive telescope (see Figure 1) and combined with a single-photon-sensitive receiver, can be used to obtain an adequate return signal. Furthermore, the combination of a much lower pulse energy with a much wider beam projected from a common transmit/receive telescope eliminates the risk of eye damage to airborne observers and therefore the need for aircraft radars and/or groundbased observers. Another option is to trade off some of this benefit to further reduce the pulse energy requirements on the laser by increasing its fire rate and averaging the measured ranges over a short time interval to create accurate "normal points". This helps to differentiate temporally correlated target returns from random photons generated by solar or receiver noise and allows the larger and more complex, higher energy, low repetition rate, modelocked lasers in conventional SLR systems to be replaced by much smaller and significantly less expensive passively Q-switched microchip lasers with repetition rates in the multi-kHz regime [12,13].

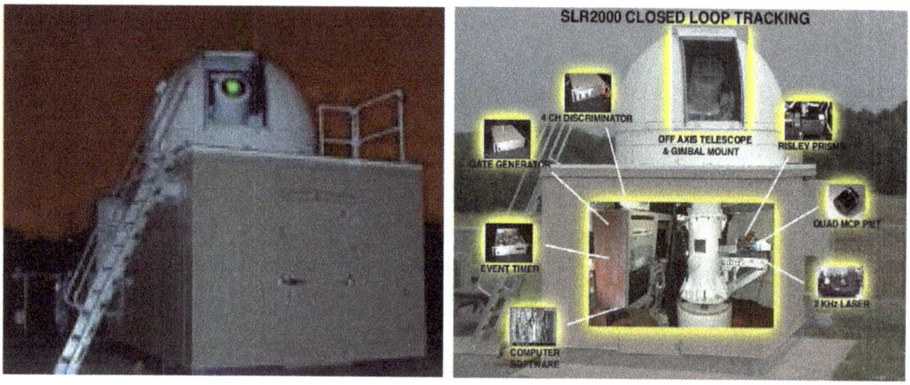

Figure 1. The prototype NASA SLR2000 system projected a 2 kHz train of low energy, 532 nm, sub-nanosecond pulses via an unobscured 30 cm primary lens in order to concentrate more photons on the satellite while simultaneously eliminating potential eye hazards.

A quadrant Multi-Channel PhotoMultiplier Tube was used to keep the received signal centered in the detector. NASA funding for the actual construction of the SLR2000 prototype began in 1997. I initially named the system "SLR2000" for two reasons: (1) the prototype was expected to be operational shortly after the start of the new millennium, and (2) the laser fire rate was increased (coincidentally) from typically less than 10 pulses per second for NASA legacy systems to 2000 pulses per second. I retired from NASA in late 2002 to join Sigma Space Corporation in Lanham, MD USA as Chief Scientist and turned over leadership of the SLR2000 program to my highly competent colleague, Jan McGarry. Later in 2003, SLR2000 successfully tracked the US TOPEX-Poseidon satellite and was renamed Next Generation Satellite Laser Ranging (NGSLR) by NASA management. NGSLR continued to operate successfully until a lightning bolt destroyed it a few years later. Three copies of an upgraded successor system, the NASA Space Geodesy Satellite Laser Ranging System (SGSLR), are expected to begin operations at three locations by 2025, i.e., GSFC in Maryland USA, the McDonald Observatory in Texas USA, and the island of Ny Alesund in Norway [9].

3. NASA Multikilohertz Microlaser Altimeter (MMLA): The First Airborne Single Photon Lidar

While examining early SLR2000 data with my NASA colleague, Jan McGarry, I noticed that a plot of measured satellite range vs predicted range exhibited a linear slope as opposed to the expected horizontal line. When I questioned Jan about the cause of the discrepancy, she replied that it was due to a time bias in the predicted orbit. This triggered the following thought: "If I can see a slope in single photon satellite data due to a time bias, why can't I see a real surface slope with a photon-counting lidar?".

NASA agreed to fund the MMLA prototype via its Instrument Incubator Program. The Civil Service team included myself, Jan McGarry, and two engineering colleagues, Thomas Zagwodzki, and Phil Dabney, along with supporting contractors from Honeywell and the newly formed Sigma Space Corporation. The MMLA or "MicroAltimeter" was largely based on SLR2000 technology and successfully measured the times-of-flight of individual photons to deduce the distances between the instrument reference and points on the underlying terrain from which the arriving photons were reflected [14]. By imaging the terrain onto a highly pixelated detector followed by a multi-channel timing receiver, one can make multiple spatially resolved range measurements of the surface via a single laser pulse. The horizontal spatial resolution is limited by the optical projection of a single pixel onto the surface. In short, a 3D image of the terrain within the laser ground spot is obtained on each laser fire, assuming at least one signal photon is recorded by each pixel/timing channel. The passively Q-switched microchip Nd:YAG laser transmitter used in MMLA measured only 2.25 mm in length and was pumped by a single 1.2 Watt GaAs laser diode. The output was frequency-doubled to take advantage of the narrower beam divergence out of the telescope as well as the higher detector counting efficiencies and narrower spectral filters available at the 532 nm wavelength. The transmitter typically produced a few microjoules of green energy in a sub nanosecond pulse at several kilohertz rates. The illuminated ground area was imaged by a diffraction-limited, off-axis telescope onto an ungated, segmented anode photomultiplier with 16 pixels (4 × 4). The effective receive aperture was about 13 cm. Each anode segment was input to one channel of a "fine" range receiver (5 cm detector-limited resolution), which recorded the times-of-flight of the individual photons. A parallel "coarse" receiver provided a lower resolution (>75 cm) histogram of atmospheric scatterers and centered the "fine" receiver gate on the previous set of returns, thereby permitting the fine receiver to lock onto ground features with no apriori range knowledge. In test flights, the system operated successfully at mid-day at aircraft altitudes up to 6.7 km (22,000 ft), with single pulse laser output energies of only a few microjoules. It also recorded kHz single photon returns from clouds, soils, man-made objects, vegetation, and water surfaces. The system also demonstrated a capability to resolve volumetrically distributed targets, such as tree canopies and the underlying terrain, and successfully performed wave height measurements and shallow water bathymetry over the Chesapeake Bay and Atlantic Ocean. The temporally correlated signal photons were reliably extracted from the random solar and/or detector noise background using an optimized post-detection algorithm.

4. Single Photon Lidar Development at Sigma Space Corporation

In early 2003, I retired from my 38-year career as a civil servant at NASA GSFC to take a position as Chief Scientist at Sigma Space Corporation, a relatively new corporation founded by Dr. Marcos Sirota and Joseph Marzouk. Sigma Space and Honeywell had provided contractor support to the MMLA Project, and Sigma was greatly interested in the further development of Single Photon Lidars (SPLs) for the commercial market. Over the next 15 years, Sigma introduced a variety of airborne single-photon-sensitive lidars, as summarized in Figure 2. The latter figure is excerpted from a comprehensive review article on Sigma's SPL technology [15].

The airborne Multiple Altimeter Beam Experimental Lidar (MABEL) at the top of Figure 2 was a 16 beam pushbroom lidar developed by Sigma for NASA as a testbed for the proposed ICESat-2 mission to be discussed in the next section. The High Altitude Lidar (HAL) and High Resolution Quantum Lidar System (HRQLS or "Hercules"), on the other hand, were scanning lidars designed for high spatial resolution, 3D imaging from high altitude aircraft. In the latter instruments, the laser beam was split into a 10 × 10 array of 100 beamlets via a dffractive element whose returns were imaged onto a matching 10 × 10 array of single photon sensitive detectors, such as a 10 × 10 segmented anode MicroChannel Plate PhotoMultiplier Tube (MCP/PMT) or a 10 × 10 Single Photon Avalanche Diode (SPAD) array.

Scanning was accomplished using the optical wedge scanner in Figure 3. A rotating single wedge traces a circle on the land surface. Over longer ranges at high altitudes, the receiver array FOV can become displaced along the circumference of the circle from the array of laser spots on the surface. To restore the overlap, we must add a "compensator wedge" which deflects the receiver FOV (or transmitter FOV but not both) at approximate right angles to the scanner wedge deflection. The Sigma Space lidars use an annular compensator wedge in which the transmitted laser beams pass unaffected through the small central hole while the receiver array FOV is angularly displaced opposite to the direction of rotation so that the detector array views the illuminated area as illustrated in Figure 3.

Figure 2. Summary of airborne single photon lidars developed at Sigma Space Corporation.

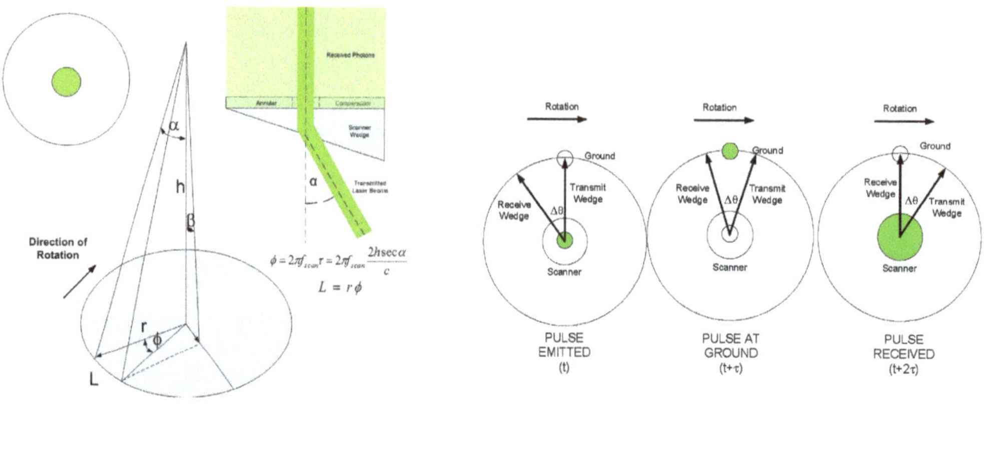

(**a**) (**b**)

Figure 3. A rotating single wedge traces a circle on the terrain. Over longer ranges, the receiver array FOV will, due to the finite speed of light, become displaced along the circumference of the circle from the array of laser spots on the surface. Therefore, we often use an annular compensator wedge in which the transmitted laser beamlets pass unaffected through a small central hole and are deflected by the optical wedge (**a**) while the receiver array FOV is angularly displaced opposite to the direction of rotation so that the detector array is viewing the illuminated area when the photons return from the surface (**b**).

Finally, photons reflected from the underlying terrain are spatially correlated but mixed with random photons due to solar illumination of the surface and/or intervening atmosphere as well as possible electronic noise in the receiver. The author previously developed an equation [16] for solar noise given by

$$\dot{n} = \left[\frac{\eta_q \eta_r}{h\nu}\right]\left[\frac{N_\lambda^0(\Delta\lambda)\Omega_r A_r}{\pi}\right]\left[\rho T^{1+sec\theta_s} cos\psi + \frac{1-T^{1+sec\theta_s}}{4(1+sec\theta_s)}\right] \quad (3)$$

where $N_\lambda{}^0 = 0.2\mathrm{W/m^2}$ Angstrom is the extraterrestrial solar irradiance impinging on the Earth's atmosphere at the 532 nm laser wavelength, η_q is the detector efficiency, η_r is the receiver optical efficiency, $h\nu$ is the laser photon energy, A_r is the area of the collecting telescope, T is the one way transmission of the atmosphere at zenith, ρ is the surface reflectance at the laser wavelength, θ_s is the solar zenith angle, ψ is the angle between the surface normal and the Sun, and $\Delta\lambda$ and Ω_r are the widths of the spectral and spatial filters. Thus, to obtain a clear 3D image of the surface, the receiver must have minimal deadtime between events and noise rejection software must be applied to the collected data. Typically, a two or three stage filter is used, i.e., one to reject the vast majority of noise relatively far from the surface followed by additional filtering to reject noise spatially intermingled with the surface returns. An example of this approach is shown in Figure 4 and final noise-edited images are provided in Figures 5–7. Figure 8 provides sample data taken over Greenland in daylight from one beamlet (#6) of the 16 beamlet airborne pushbroom lidar MABEL The latter systemwas built to simulate the original 16 beamlet lidar proposed as a demonstration mission for ICESat-2. The figure clearly shows the ability of the lidar to distinguish the terrain from the background solar radiation over a wide range of slopes.

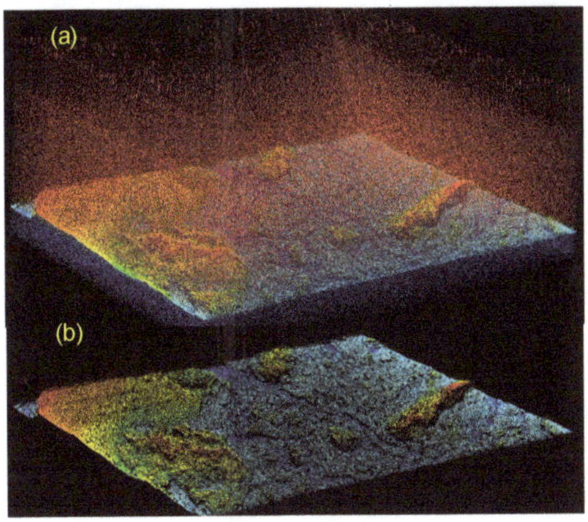

Figure 4. Editing Noise (**a**) Unedited point cloud of Greenland terrain (surface reflectivity > 0.9) shows a fair amount of solar noise above (red haze) and below (blue haze) the surface. (**b**) same point cloud image after use of Sigma-developed noise-editing filters.

The success of the high altitude airborne SPLs, designed and manufactured by Sigma Space Corporation, clearly demonstrated the amazing detail that could be obtained over a wide variety of landscapes. These included rural and forested areas [17], tall city buildings, homes, and other manmade structures, rugged and sloped terrain, ice floes, power lines, and the surface and bottoms of shallow water bodies with depths up to several tens of meters.

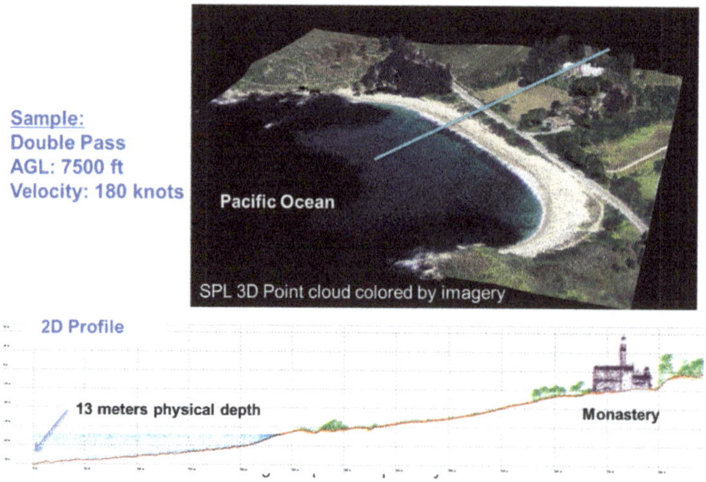

Figure 5. (**Top**) Lidar image of the Pacific coastline in Port Lobos, California. (**Bottom**) Detailed look along the blue line in the top figure showing a hilltop monastery, the heights of various trees, the surface of the Pacific Ocean, and the ocean bottom to a depth of 13 m (42.7 ft).

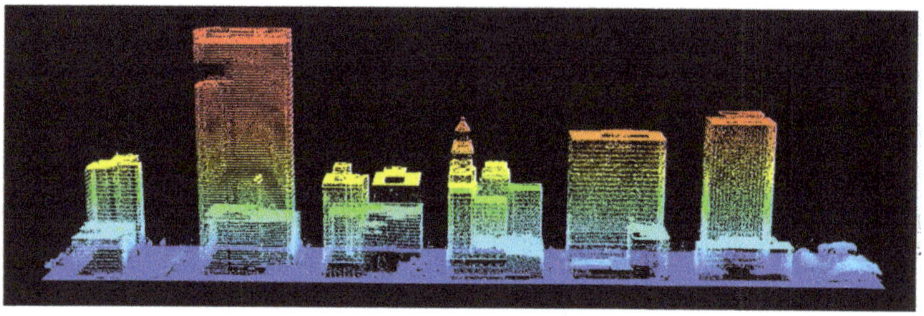

Figure 6. Single photon lidar image of downtown Houston, Texas, taken by the Sigma HRQLS system (pronounced "Hercules"). Colors were arbitrarily assigned based on height above the ground.

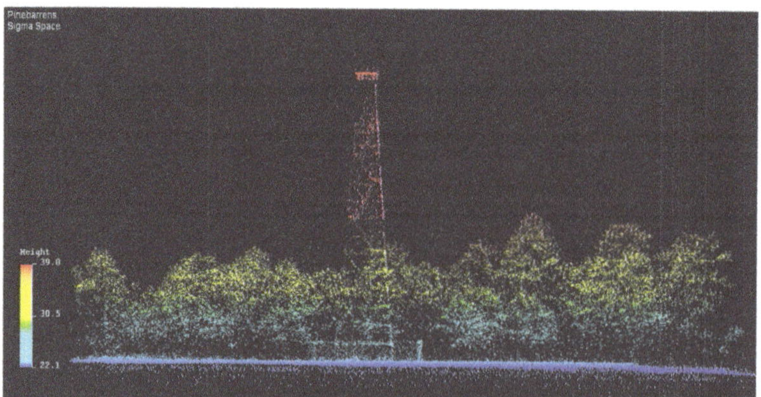

Figure 7. Side view of an airborne 3D image of a fire tower surrounded by a chain link fence and trees of varying height. Colors are arbitrarily assigned based on height above the surface.

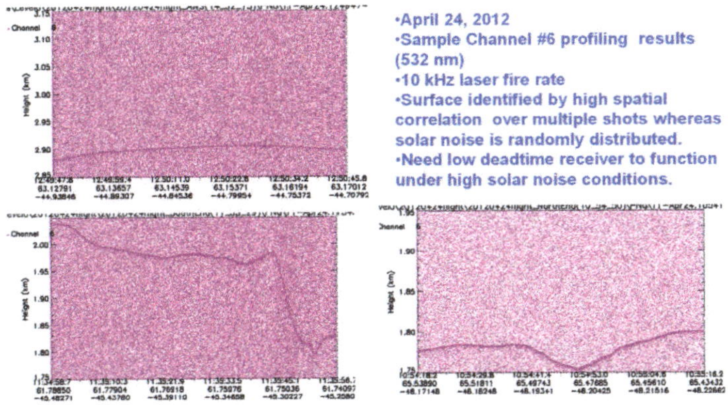

Figure 8. Sample surface data from a single channel (#6) of the 16-channel airborne MABELpush-broom lidar taken over the Greenland ice sheet from an altitude of 20 km [18]. The data demonstrated the lidar's ability to observe a wide range of surface slopes in the presence of high-intensity solar (or detector) noise due to high spatial correlation of the surface counts and a low deadtime receiver.

5. Configuring the ICESat-2 Single Photon Lidar

NASA had previously flown multiple spaceborne multiphoton lidars, including the following:

- the Mars Orbiter Laser Altimeter (MOLA) from 1997 to 2001;
- the GLAS (Geoscience Laser Altimeter System) on ICESat-1 about the Earth from 2003 to 2009;
  - the MESSENGER mission to Mercury from 2011 to 2015;
  - and the Lunar Observer Laser Altimeter (LOLA) about the Moon from 2009 to the present.

The aforementioned multiphoton spaceborne lidars operated at a few tens of pulses per second, and it rapidly became clear to the author that spaceborne SPL technology could be more lightweight, less complex, and provide orders of magnitude more surface data, especially in orbit about planets or moons with less atmosphere and surface diversity than Earth [16].Therefore, in 2006, I gave several presentations to the NASA ICESat-2 Project team at Goddard Space Flight Center. The presentations outlined the potential benefits of a multibeam, single photon lidar in Earth orbit. The original design presented a single 5 Watt, actively Q-switched, frequency doubled, subnanosecond pulse Nd:YAG microlaser operating at the 532 nm wavelength and generating a train of 500 microjoule pulses at a rate of 10 kHz. The beam would then be split into 16 equally spaced, 310 mW beamlets distributed along a line perpendicular to the satellite flight path. Thus, the proposed configuration was designed to generate 160,000 surface measurements per second compared to only 40 measurements per second for the predecessor GLAS single beam, multiphoton lidar flown on ICESat-1. In addition, the green wavelength would permit some degree of penetration in water bodies for bathymetric measurements.

Initially, the ICESat-2 project team saw potential merit in the SPL approach and proposed to fly a copy of the earlier GLAS system along with an experimental version of the proposed 16 beam SPL on the ICESat-2 mission. However, budget problems later forced the Project to abandon the dual lidar concept. Fortunately, results from the NASA-funded airborne version of Sigma's 16 beam pushbroom SPL, MABEL, helped to bolster confidence in the technique as illustrated in Figure 8 [18]. Nevertheless, it was not yet entirely clear that the 16 beam lidar proposed by Sigma could meet all of the science requirements of the mission.

Ultimately, the Project chose a modified version of my proposed single photon lidar which consisted of 3 "strong" and 3 "weak" beams for reasons to be described shortly. The strong beams would each contain roughly four times the pulse energy (2 mJ) of one of the original 16 weak beams thereby reducing the surface sampling rate from 160,000 measurements per second to 60,000 per second but still representing a factor of 1500 increase in potential surface detection rate relative to the original GLAS system on ICESat-1.

In order to reach this final instrument configuration, I again collaborated with my longtime NASA colleague and subsequent ICESat-2 Algorithm Lead, Jan McGarry, to assess the performance of various beam strengths for each of the 13 measurement scenarios in Table 1. I devised a Differential Cell Count (DCC) approach while Jan suggested a Noise Only (NO) method for extracting the weak surface signal from solar background noise over the range of measurement scenarios which were considered critical by the ICESat-2 Project Science Team. As outlined in Table 1, these scenarios included cryosphere, biomass, and ocean measurements. Furthermore, successful measurements were desired for terrain slopes up to 45 degrees. In the process, we each used lidar link analyses and solar noise models appropriate for each of the 13 measurement scenarios provided by the science team and optimized the size of the range bin. The number of beamlets and their powers were also varied, keeping the overall laser power constant, in order to determine their compatibility with the measurement goals.

Table 1. The ATLAS science team identified 13 measurement scenarios crucial to their overall mission, including six ice-related scenarios (1a through 6a), three tree penetration scenarios (8a, 9a, and 10a), and three wind measurements (10a, 10b, and 10c). The expected signal counts and noise rates for each scenario are listed in the final two columns of the table.

GSFC Design Case	Range Bin, ρ_b (m)	Range Window ΔR (km)	Description	Signal Counts, n_s Weak beam	Noise Rate (MHz)
1a	30	6	Ice Sheet, Interior Winter	2.15	0.5
2c	30	6	Ice Sheet, Interior, Summer, Heavy Snow	0.43	6.0
3c	30	6	Outlet Glacier, Winter, Heavy Snow	0.15	0.50
4	30	6	Outlet Glacier, Summer	0.68	4.69
5b	12	6	Ice Lead, Winter	0.26	0.5
6a	12	6	Ice, Summer	0.6	5.73
7a min	12	6	Tropical Flat, min leaf	0.196	3.37
7a max	12	6	Tropical Flat, max leaf	0.154	2.73
8a	30	6	Temperate Flat	0.44	1.21
9a	30	6	Boreal Flat	0.324	1.47
10a	12	1	Conical Scan, Low Wind	0.49	2.68
10b	12	1	Conical Scan, Moderate Wind	0.21	1.94
10c	12	1	Conical Scan, High Wind	0.122	1.71

Table 2 is extracted from a PowerPoint presentation by the author to the ICESat-2 science team in December 2011. As can be seen from the far left column in Table 2, four different combinations of beam strengths were considered: (1) 16 equal weak beams (as originally proposed by the author for the test flight configuration); (2) 6 equal beams each containing 2.5 times the energy in the original weak beam; (3) 5 equal beams each containing 3 times the energy in the weak beam; and (4) strong beams each containing 4 times the energy in the weak beam. In turn, each of these categories were further divided into two elements depending on which detection algorithm was being used, i.e., the Differential Cell Count (DCC) or Noise Only (NO) method. Dark blue color indicates excellent performance, dark green signifies very good, with light green, yellow, orange, and red indicating increasingly degraded performance in each measurement category. To summarize, the three strong beams perform well in all of the measurement scenarios while the remaining 3 weak beams can double the measurement density in certain science categories such as 1a, 8a, and 10a. The above conclusions were considered conservative by the Science Team since the calculations assumed, for simplicity, a worst case where the surface photons are distributed uniformly in time by either a constant slope within the frame or the tree canopy which they are not.

Table 2. Summary of beam strength performance for the 13 ICESat-2 science goals.

Beam Strength	Science	Cryosphere						Biomass				Ocean Cal		
	Range Window	6 km						6 km				1 km		
	Design Case	1a	2c	3c	4	5b	6a	7a max	7a min	8a	9a	10a	10b	10c
	Bin Size (m)	18	18	18	18	12	12	30	30	30	30	12	12	12
Weak 1X	DCC	>45° >141 m	<7° <17 m	<10° <25 m	<26° <68 m	<26° <68 m	<16° <17 m	0° 0 m	0° 0 m	<41° <122 m	<22° <56 m	<28° <77 m	<8° <21 m	<2° <4 m
	NO	>45° >141 m	<6° <15 m	<12° <30 m	<25° <65 m	<33° <93 m	<16° <17 m	0° 0 m	0° 0 m	<34° <96 m	<19° <48 m	<29° <80 m	<7° <17 m	<2° <4 m
6 Equal 2.5X	DCC	>45° >141 m	<39° <117 m	>45° >141 m	>45° >141 m	>45° >141 m	<43° <136 m	<16° <40 m	<23° <60 m	>45° >141 m	>45° >141 m	>45° >141 m	<37° <105 m	<20° <51 m
	NO	>45° >141 m	<34° <96 m	>45° >141 m	>45° >141 m	>45° >141 m	>45° >141 m	<12° <30 m	<18° <47 m	>45° >141 m	>45° >141 m	>45° >141 m	<40° <118 m	<20° <52 m
5 Equal 3X	DCC	>45° >141 m	>45° >141 m	>45° >141 m	>45° >141 m	>45° >141 m	>45° >141 m	<24° <63 m	<31° <85 m	>45° >141 m	>45° >141 m	>45° >141 m	<43° <135 m	<26° <68 m
	NO	>45° >141 m	>45° >141 m	>45° >141 m	>45° >141 m	>45° >141 m	>45° >141 m	<21° <53 m	<28° <77 m	>45° >141 m	>45° >141 m	>45° >141 m	<40° <118 m	<27° <71 m
Strong 4X	DCC	>45° >141 m	>45° >141 m	>45° >141 m	>45° >141 m	>45° >141 m	>45° >141 m	<37° <105 m	<44° <138 m	>45° >141 m	>45° >141 m	>45° >141 m	>45° >141 m	<37° <105 m
	NO	>45° >141 m	>45° >141 m	>45° >141 m	>45° >141 m	>45° >141 m	>45° >141 m	<36° <103 m	<43° <135 m	>45° >141 m	>45° >141 m	>45° >141 m	>45° >141 m	<40° <118 m

Color Code for Cryosphere
Red: Inadequate—Fails to meet minimum 4-degree slope requirement.
Orange: Marginal—Can handle slopes/relief up to 10°/25 m.
Yellow: Fair—Can handle slopes/relief greater than 10°/25 m.
Light Green: Good—Can handle slopes/relief greater than 20°/52 m.
Dark Green: Very Good—Can handle slopes/relief greater than 30°/82 m.
Dark Blue: Outstanding—Can handle slopes/relief in excess of 40°/118 m.

Color Code for Biomass
Red: Inadequate—Cannot detect minimum tree height stands on 4-degree slope.
Orange: Marginal—Can detect minimum tree height stands on 4-degree slope. Total slope/relief = 11°/28m for 7a, 10°/25m for 8a, and 16°/40m for 9a.
Yellow: Fair—Can detect average tree height stands on 4-degree slope. Total slope/relief = 15°/38 m for 7a, 12°/30m for 8a, and 20°/52m for 9a.
Light Green: Good—Can detect maximum tree height stands on 4-degree slope. Total slope/relief = 20°/52 m for 7a, 14°/38m for 8a, and 23°/60m for 9a.
Dark Green: Very Good—Can detect maximum tree height stands on 14-degree slope. Total slope/relief = 30°/82m for 7a, 24°/64m for 8a, and 33°/93m for 9a.
Dark Blue: Outstanding—Can detect maximum tree height stands on 24+ degree slope. Total slope/relief = 40°/118m for 7a, 34°/97m for 8a, and 43°/137m for 9a.

Color Code for Ocean Cal
Red: Inadequate—Cannot detect ocean surface with a 4-degree cone half angle.
Orange: Marginal—Can detect ocean surface with conical scan angles between 4 and 5 degrees.
Yellow: Fair—Can detect ocean surface with conical scan half angles between 5 and 7 degrees.
Light Green: Good—Can detect ocean surface with conical scan half angles between 7 and 10 degrees.
Dark Green: Very Good—Can detect ocean surface with conical scan in excess of 10 degrees.

6. Summary

As described herein, the precision ranging capability of the current ATLAS lidar relied heavily on almost four decades of relevant experience in Satellite Laser Ranging (SLR) during which the ranging precision improved from 2 m to a few millimeters. Subsequently, the growing international science and engineering demands for precise satellite ranging at a lower manufacturing and operational cost resulted in NASA's development of the single photon sensitive SLR2000 system. The latter design drew heavily from early NASA laser space communication studies of optical antenna gain conducted in the 1970s. These studies advocated for the use of a single telescope to (1) transmit the pulse in a significantly less divergent beam (thereby placing an orders of magnitude higher fraction of the transmitted laser pulse energy on the target), and (2) capturing some of the energy reflected by the satellite retroreflectors. This innovation not only significantly reduced the size and cost of the optical telescope, tracking mount, and shelter, but it also greatly reduced the required laser pulse energy. Thus, the large, complex, expensive, and low repetition rate mode-locked lasers could be replaced by relatively inexpensive, compact, low energy, actively or passively Q-switched microchip lasers producing tens of thousands of subnanosecond pulses per second.

In this paper, the author has attempted to describe for the reader the highly complicated technical path which ultimately led NASA researchers and their contractors to the highly successful ATLAS Multibeam Single Photon Lidar (SPL). The latter instrument was launched, in September 2018, aboard the ICESat-2 spacecraft into a roughly 500 km high orbit about the Earth. At that altitude, the six beams and 10 kHz laser pulse rate permit a total of 60,000 surface measurements per second and surface measurements every 70 cm along the flight path for each of the 6 beamlets. The ICESat-2 science goals included few cm accuracy topographic measurements over land, ice, and sea surfaces as well as tree canopy and cloud heights.

The ATLAS lidar includes an 80 cm diameter telescope which can redirect the beams. An initial evaluation of ATLAS performance published in 2021, after more than one trillion surface measurements had been made, concluded with the following statement. "The ICESat-2 geolocated photons show a horizontal position accuracy of 3.6 m (1 σ) over both long length scales and through local validation, and a vertical accuracy of better than 10 cm" [19]. A wide range of publications reporting on the ICESat-2 instrument and data analyses can be found on a routinely updated NASA website at https://icesat-2.gsfc.nasa.gov (accessed on 1 September 2024).

Funding: This research received no external funding.

Data Availability Statement: The original contributions presented in the study are included in the article, further inquiries can be directed to the corresponding author.

Conflicts of Interest: The authors declare no conflict of interest.

References

1. Maiman, T.H. Optical Radiation in Ruby. *Nature* **1960**, *187*, 493–494. [CrossRef]
2. Plotkin, H.H.; Johnson, T.S.; Spadin, P.; Moye, J. Reflection of Ruby Laser Radiation from Explorer XXII. *Proc. IEEE* **1965**, *53*, 301–302. [CrossRef]
3. Alley, C.O.; Chang, R.; Currie, D.; Mullendore, J.; Poltney, S.; Rayner, J.D.; Silverberg, E.; Steggerda, C.; Plotkin, H.; Williams, W.; et al. Apollo 11 Laser Ranging Retroreflector: Initial Measurements from the McDonald Observatory. *Science* **1970**, *167*, 368–370. [CrossRef]
4. Degnan, J.J. Millimeter Accuracy Satellite Laser Ranging: A Review. *Contrib. Space Geod.Geodyn. Technol.* **1993**, *25*, 133–162.
5. Degnan, J.J. A Tutorial on Retroreflectors and Arrays Used in Satellite and Lunar Laser Ranging. *Photonics* **2023**, *10*, 1215. [CrossRef]
6. Nicolas, J.; Pierron, F.; Kasser, M.; Exertier, P.; Bonnefond, P.; Barlier, F.; Haase, J. French transportable laser ranging station: Scientific objectives, technical features, and performance. *Appl. Opt.* **2000**, *39*, 402–410. [CrossRef] [PubMed]
7. Pearlman, M.R.; Noll, C.E.; Pavlis, E.C.; Lemoine, F.G.; Combrinck, L.; Degnan, J.J.; Kirschner, G.; Schreiber, K.U. The ILRS: Approaching 20 Years and Planning for the Future. *J. Geod.* **2019**, *93*, 2161–2180. [CrossRef]
8. Wilkinson, M.; Schreiber, K.U.; Prochazka, I.; Moore, C.; Degnan, J.J.; Kirchner, G.; Zhonping, Z.; Dunn, P.; Shargorodsky, V.; Sadovnikov, M.; et al. The Next Generation of Satellite Laser Ranging Systems. *J. Geod.* **2019**, *93*, 2227–2247. [CrossRef]
9. McGarry, J.F.; Degnan, J.J.; Cheek, J.W.; Clarke, C.B.; Diegel, F.; Donovan, H.L.; Horvath, J.E.; Marzouk, M.; Nelson, A.R.; Patterson, D.S.; et al. NASA's Satellite Laser Ranging Systems for the 21st Century. *J. Geod.* **2019**, *93*, 2249–2262. [CrossRef]
10. Klein, B.J.; Degnan, J.J. Optical Antenna Gain 1: Transmitting Antennas. *Appl. Opt.* **1974**, *13*, 2134–2141. [CrossRef] [PubMed]
11. Degnan, J.J.; Klein, B.J. Optical Antenna Gain 2: Receiving Antennas. *Appl. Opt.* **1974**, *13*, 2397–2401, Erratum in *Appl. Opt.* **1974**, *13*, 2762. [CrossRef] [PubMed]
12. Degnan, J.J. Optimal Design of Passively Q-switched Microlaser Transmitters for Satellite Laser Ranging. In Proceedings of the 10th International Workshop on Laser Ranging, Shanghai, China, 11–15 November 1996; pp. 334–343.
13. Degnan, J.J.; Zayhowski, J.J. SLR2000 Microlaser Performance: Theory and Experiment. In Proceedings of the 11th International Workshop on Laser Ranging, Deggendorf, Germany, 21–25 September 1998; pp. 453–468.
14. Degnan, J.J.; McGarry, J.F.; Dabney, P.; Zagwodzki, T.W. Design and Performance of an Airborne Multi-kilohertz Photon-Counting Microlaser Altimeter. *Int. Arch. Photogramm. Remote Sens. Spat. Inf. Sci.* **2001**, *34*, 9–16.
15. Degnan, J.J. Scanning, multibeam, single photon lidars for rapid, large scale, high resolution topographic and bathymetric mapping. *Remote Sens.* **2016**, *8*, 958. [CrossRef]
16. Swatantran, A.; Tang, H.; Barrett, T.; DeCola, P.; Dubayah, R. High-Resolution Forest Structure and Terrain Mapping over Large Areas using Single Photon Lidar. *Sci. Rep.* **2016**, *6*, 28277. [CrossRef] [PubMed]
17. Degnan, J.J. Photon-counting multikilohertz microlaser altimeters for airborne and spaceborne topographic measurements. *J. Geodyn.* **2002**, *34*, 503–549. [CrossRef]
18. McGill, M.; Markus, T.; Scott, V.S.; Neumann, T. The Multiple Altimeter Beam Experimental Lidar (MABEL): An Airborne Simulator for the ICESat-2 Mission. *J. Atmos. Ocean. Technol.* **2013**, *30*, 345–352. [CrossRef]
19. MacGrudor, L.; Neumann, T.; Kurtz, N. ICESat-2 Early Mission Synopsis and Observatory Performance. *Earth Space Sci.* **2021**, *8*, 5. [CrossRef] [PubMed]

photonics

MDPI

Article

Real-Time Simulation of Clear Sky Background Radiation in Gas Infrared Remote Sensing Monitoring

Shengquan Shu [1,2], Jianguo Liu [1,2], Liang Xu [2,*], Yuhao Wang [1,2], Yasong Deng [1,2] and Yongfeng Sun [2]

1 School of Environmental Science and Optoelectronic Technology, University of Science and Technology of China, Hefei 230026, China; sqshu@aiofm.ac.cn (S.S.); jgliu@aiofm.ac.cn (J.L.); wangyuhao@aiofm.ac.cn (Y.W.); ysdeng@aiofm.ac.cn (Y.D.)

2 Key Laboratory of Environmental Optics and Technology, Anhui Institute of Optics and Fine Mechanics, Hefei Institutes of Physical Science, Chinese Academy of Sciences, Hefei 230031, China; yfsun@aiofm.ac.cn

* Correspondence: xuliang@aiofm.ac.cn

Abstract: During the process of infrared remote sensing monitoring, obtaining real-time measurements of sky background radiation is extremely inconvenient. The current methods incur a certain amount of lag. In this study, within the existing theoretical framework, a fast transmittance calculation method using interpolation was adopted, and a simplified transmission model was established. This led to the development of a new and simplified method for rapid temperature and humidity retrieval. Compared to the line-by-line integration method, the interpolation method significantly improves the speed of transmittance calculation by several tens of times, while maintaining a high level of accuracy. The relative deviation between the results obtained using the interpolation method and those obtained through line-by-line integration is less than 1 ‱. With the proposed method, temperature and humidity profile information can be retrieved from measured spectra within 5 min and corresponding background spectra can be obtained. The differences between the calculated background radiation and the measured spectra using the new method are smaller, making it more suitable for calculating sky background radiation. Additionally, the rapid retrieval results of the temperature profiles in the lower atmosphere have a certain level of accuracy (the mean deviation is less than 2 K).

Keywords: infrared remote sensing; sky background radiation; ground based; temperature and humidity profile

1. Introduction

Fossil fuels continue to be the most crucial and primary energy sources in use today [1]. Throughout the extraction, production, and refining processes of these fuels and their derivatives, companies associated with fossil energy may emit and leak toxic and harmful gases, resulting in air pollution and serious safety incidents [2]. Consequently, air monitoring has become an indispensable part of the production process. Traditional monitoring methods primarily involve deploying monitoring equipment at emission sources for continuous monitoring and manually collecting samples from various locations. Although these methods offer high precision, the complexity of the gas compositions poses challenges, making the process time-consuming and labor-intensive. Furthermore, traditional methods are often inadequate in effectively monitoring unorganized emissions, such as pipeline leaks, which frequently serve as the primary cause of production safety accidents [3,4].

Fourier Transform Infrared (FTIR) spectroscopy technology is a non-contact technique that allows for the simultaneous analysis of multiple components online. When combined with passive scanning remote sensing technology, it enables comprehensive and simultaneous monitoring of various gas components in factory areas. This method represents a highly effective means of monitoring unorganized emissions, such as pipeline leaks, and has already been applied in a commercialized manner [5–9].

Using passive Fourier Transform Infrared (FT-IR) spectroscopy is typically a semi-quantitative method for leak monitoring [10]. By comparing the real-time monitoring spectra with background spectra, we can analyze the general situation of gas leaks (Figure 1a). Therefore, obtaining accurate background spectra is crucial for this monitoring process [11]. The monitoring background is usually divided into two categories: ground object background (Figure 1b) and sky background (Figure 1c). The ground object background typically consists of buildings or tall, dense vegetation, which generally do not change significantly over a prolonged monitoring period. Thus, it suffices to obtain this background at the beginning of the measurement process. In contrast, the sky background can be classified into cloudy and clear conditions. The sky background radiation in clear sky exhibits significant changes because of daily fluctuations in atmospheric conditions (primarily temperature and humidity). This requires frequent measurements of the sky background, which can be quite inconvenient for the overall monitoring process. To address the challenges associated with monitoring sky background, atmospheric profile parameters can be utilized along with mathematical models for simulation calculations [12]. While under a cloudy sky, the factors affecting the background spectrum become more complex in the presence of clouds, and from an engineering application standpoint, there is no well-established method to address this issue yet. The way we currently deal with clouds is that we ideally assume that cloud conditions remain the same for a certain period of time and then periodically collect the sky background when there are clouds.

(a) (b) (c)

Figure 1. Three-layer radiation transmission model and different monitoring backgrounds: (**a**) three-layer radiation transmission model; (**b**) ground object background; (**c**) sky background.

Atmospheric parameter profiles can be obtained from many datasets. Most atmospheric components exhibit minimal variation, allowing for calculations based on average values. However, for parameters such as temperature and humidity, which exhibit significant daily fluctuations, more recent results are necessary for more accurate calculations of sky background radiation. Taking the ERA5 hourly pressure-level dataset provided by the European Centre for Medium-Range Weather Forecasts (ECMWF) as an example, this dataset is updated daily but has a lag of about 5 days. Thus, we wondered whether it is possible to extract temperature and humidity profiles in real-time from monitoring spectra for background calculations. In fact, obtaining temperature and humidity profiles from ground-based infrared radiation spectra has already become a reality, with dedicated ground-based infrared radiometers [13,14] and corresponding algorithms for temperature and humidity monitoring [15–18]. However, scientific research demands very high monitoring precision, necessitating stricter requirements for analysis algorithms, greater computational load, and longer processing times. Therefore, the aim of this paper is to simplify the existing technologies and algorithms to achieve rapid measurement of temperature and humidity profiles based on ground-based infrared radiation spectra, thereby facilitating background radiation calculations and further applications in gas leak monitoring.

2. Materials and Methods

2.1. Retrieval Strategy

The retrieval of temperature and humidity profiles involves reverse calculations of the atmospheric state using observed values and transfer models, necessitating the solution of an implicit equation [19]. Thus, it is necessary to find the most suitable solution through iterations. In this paper, we employ the Newton iteration method, which is generally described as follows:

$$\vec{x}_{i+1} = \vec{x}_i - \left[\nabla^2 J\left(\vec{x}_i\right)\right]^{-1} \nabla J\left(\vec{x}_i\right), \quad (1)$$

$J\left(\vec{x}_i\right)$ is the cost function, ∇ represents the gradient symbol, $\left[\nabla^2 J\left(\vec{x}_i\right)\right]$ also known as the Hessian matrix.

In this study, we utilized the maximum probability cost function and applied the Levenberg-Marquardt method to modify the Hessian matrix. As a result, we derived the following specific iteration formula:

$$\vec{x}_{i+1} = \vec{x}_i - \left[S_a^{-1} + K_i^T S_\varepsilon^{-1} K_i + u_i \boldsymbol{I}\right]^{-1} \left\{S_a^{-1}\left(\vec{x}_i, -, \vec{x}_a\right) - K_i^T S_\varepsilon^{-1}\left[\vec{y} - F\left(\vec{x}_i\right)\right]\right\}, \quad (2)$$

$\vec{x}_0 = \vec{x}_a$, $\vec{x}_a$ is the mean value of the prior profile, and $\vec{y}$ is the observed radiance. $F\left(\vec{x}_i\right)$ means the radiance calculated by the transfer model when the state vector is $\vec{x}_i$. S_a and S_ε represent the prior covariance matrix and error covariance matrix, respectively.

S_a is calculated directly from prior information, whereas the S_ε employs a diagonal matrix with the diagonal elements derived from the mean noise of the measured spectral data [16]. In this study, the diagonal values of the S_ε are set to 0.1. The Jacobian matrix K_i is recalculated for each iteration using the finite differences method. $\boldsymbol{I}$ is a unit matrix and u_i represents the L-M parameter. The superscripts -1 and T represent the inverse and transpose of the matrix, respectively.

In this paper, the L-M parameter is set based on empirical knowledge. The method employed in this study is to set $u_i = p_i + q$. If the minimum eigenvalue of the Hessian matrix, denoted as σ_i, is not less than zero, we $p_i = 0$; otherwise, we set $p_i = |\sigma_i|$. For temperature retrieval, $q = 30$, while for humidity, $q = 10^{-5}$.

The iteration process terminates when the number of iterations exceeds 10 ($i > 10$) or when the difference between the results of two consecutive iterations falls below a specified threshold (Equation (3)).

$$\left\|\vec{x}_{i+1} - \vec{x}_i\right\|_2 \leq 1, \quad (3)$$

2.2. Vertical Grid

The vertical pressure grid used is shown in Table 1, which is consistent with the grid utilized in the Smith study [17], featuring a maximum of 50 layers. In an actual retrieval process, the lowest layer is determined based on the surface atmospheric pressure. For example, if the surface pressure is 992 hPa, the lowest layer pressure is set to 990 hPa. Conversely, if the surface pressure is 988 hPa, the lowest layer pressure is set to 980 hPa, and so on.

Table 1. Vertical pressure (in hPa) grid used in profile retrieval.

1000	950	900	800	700	600	350	150	30	7
990	940	880	780	680	550	300	125	25	5
980	930	860	760	660	500	250	100	20	4
970	920	840	740	640	450	200	75	15	3
960	910	820	720	620	400	175	50	10	2

2.3. Prior Information

In scientific research involving temperature and humidity profile observations, equipment is typically installed at fixed observing stations. These stations regularly release radiosondes to obtain high-precision and high-vertical resolution profiles as prior information. However, it is impractical for operational applications. Therefore, we rely on the ERA5 dataset as a substitute for prior information. The ERA5 dataset is a global atmospheric reanalysis dataset that provides a comprehensive set of atmospheric variables, including temperature and humidity profiles. These profiles are derived from a combination of observations and numerical models, making the ERA5 dataset a reliable source of prior information for temperature and humidity retrieval in operational applications.

The dataset includes information across 37 pressure levels, ranging from 1000 hPa to 1 hPa, covering the entire globe with a horizontal spatial resolution of up to $0.25° \times 0.25°$. The prior information consists of two types of water vapor concentration data: relative humidity (RH, %) and the mass mixing ratio ($C_{m,1}$, kg/kg), as well as the ozone mass mixing ratio ($C_{m,2}$, kg/kg). Since the concentration inputs in the transfer model are in the volume mixing ratio, a unit conversion is required.

For water vapor, the saturated vapor pressure E (hPa) at a given temperature is calculated using the Goff–Gratch equation. When the temperature t is below 273.15 K, the saturated vapor pressure is as follows [20]:

$$\lg E = k_1 \times \left(\frac{273.16}{t} - 1\right) + k_2 \times \lg\left(\frac{273.16}{t}\right) + k_3 \times \left(1 - \frac{t}{273.16}\right) + \lg k_4, \quad (4)$$

Among them, $k_1 = -9.09718$, $k_2 = -3.56654$, $k_3 = 0.876793$, and $k_4 = 6.1071$.

When the temperature t is above 273.15 K, the formula is as follows:

$$\lg E = k_5 \times \left(\frac{373.16}{t} - 1\right) + k_6 \times \lg\left(\frac{373.16}{t}\right) + k_7 \times \left(10^{k_9 \times (1 - \frac{t}{373.16})} - 1\right) + k_8 \times \left(10^{k_{10} \times (\frac{373.16}{t} - 1)} - 1\right) + \lg k_{11} \quad (5)$$

Among them, $k_5 = -7.90298$, $k_6 = 5.02808$, $k_7 = -1.3816 \times 10^{-7}$, $k_8 = 8.1328 \times 10^{-3}$, $k_9 = 11.344$, $k_{10} = -3.49149$, and $k_{11} = 1013.246$.

Then, we can calculate the volume mixing ratio from E:

$$C_{v,1} = 10^4 \times RH \times E / P_{air}, \quad (6)$$

$C_{v,1}$ represents the volume mixing ratio of water vapor, ppmv. P_{air} is the air pressure, hPa.

For ozone, we need the average relative molecular mass of air, M_{air}. It can be calculated using the mass mixing ratio and volume mixing ratio of water vapor, or alternatively, empirical values can be directly utilized:

$$M_{air} = C_{v,1} \times M_1 / \left(10^6 \times C_{m,1}\right) \text{ or } M_{air} = 29, \quad (7)$$

M_1 is the relative molecular mass of water.

Then, we get the volume mixing ratio of ozone, $C_{v,2}$, as follows:

$$C_{v,2} = 10^6 \times M_{air} \times C_{m,2} / M_2, \quad (8)$$

M_2 is the relative molecular weight of ozone.

In this paper, we selected data from four time periods (0000 UTC, 0600 UTC, 1200 UTC, and 1600 UTC) in April each year from 2019 to 2023 in Hefei (31.9~32.1° N, 117.1~117.3° E) as prior information. A linear interpolation is required to match the prior information to the vertical pressure grid.

2.4. Spectrum and Radiation Transfer Model

2.4.1. Spectrum to Be Retrieved

The purpose of this study is to explore the feasibility of real-time acquisition of sky background radiation. Additionally, assessing the quality of the measured spectra is challenging due to potential errors introduced by instrument performance and calibration accuracy. To mitigate these errors, simulated spectra derived from gas profiles and radiation transfer models are utilized as the spectra of interest for retrieval in this study.

2.4.2. Simplified Radiation Transfer Model

Most major atmospheric components have a certain background concentration, and most trace constituents only exhibit significant absorption in the spectrum when observed at low elevation angles. Therefore, in the radiation transfer model used in this study, only 12 gases included in the US Standard Atmosphere model are considered. Before calculating the downwelling radiation, we need to make two assumptions [21]:

1. In the radiation transfer model, the atmosphere is assumed to be in thermodynamic equilibrium, allowing us to use the Planck intensity as the source function in accordance with Kirchhoff's law.
2. The atmosphere is considered to be plane-parallel, implying that the radiance or intensity varies only in the vertical direction.

When only absorption and emission of radiation are considered, a simplified radiation transfer model can be written in the form:

$$I^{\downarrow}_{\nu,\theta} = \sum_{i=1}^{N} B_{\nu,i} \times (1 - \tau_{\nu,\theta,i}) \times \tau^{*}_{\nu,\theta,i} \quad \tau^{*}_{\nu,\theta,i} = \prod_{j=1}^{i} \tau_{\nu,\theta,j-1}, \tag{9}$$

In the equation, N represents the number of atmospheric layers; $B_{\nu,i}$ indicates the blackbody radiation intensity at wavenumber ν for the i-th layer of the atmosphere at a given temperature; $\tau_{\nu,\theta,i}$ denotes the transmittance along the observation direction at wavenumber ν for the i-th atmospheric layer, $\tau_{\nu,\theta,0} = 1$; $\tau^{*}_{\nu,\theta,i}$ signifies the total transmittance along the observation direction at that wavenumber for all atmospheric layers below the i-th layer; and θ is the zenith Angle.

When calculating the entire layer of atmospheric downwelling radiation at the surface, the blackbody radiation intensity and emissivity $(1 - \tau_{\nu,\theta,i})$ of each layer are calculated first, and thus the radiation intensity $B_{\nu,i} \times (1 - \tau_{\nu,\theta,i})$ of each layer is obtained. When the radiation of this layer is absorbed by the atmosphere with transmittance $\tau^{*}_{\nu,\theta,i}$, the radiation reaching the surface is $B_{\nu,i} \times (1 - \tau_{\nu,\theta,i}) \times \tau^{*}_{\nu,\theta,i}$, and the contribution of each layer can be accumulated to obtain the entire layer of atmospheric downwelling radiation at the surface.

2.4.3. Fast Transmittance Calculation

To calculate the transmittance in the radiation transfer model, we employ a simple method based on line-by-line integration. We assume that, at the same pressure, the absorption coefficient of gases at a specific wavelength has a linear relationship with temperature within a narrow range. To create a database of line-by-line absorption coefficients for various gases at different temperatures and pressures, we utilized the HAPI program [22]. In this study, the pressure conditions are consistent with the fixed vertical grid mentioned above, with a temperature range of 200 K to 320 K, at intervals of 0.5 K, and a wavenumber step of 0.1 cm^{-1}. The maximum step size recommended by the HAPI program is 0.1 cm^{-1}; a step size larger than this value may affect computational accuracy. There are two principles regarding the setting of the step size: (1) ensure sufficient computational accuracy, and (2) ensure that the absorption database generated from this step size does not consume excessive memory. We used the HAPI program to calculate the line-by-line absorption coefficients of 12 gases under the specified conditions, thereby establishing a database. The use of the HAPI program was strictly performed according to the manual. The program and manual can be accessed at the following URL: https://hitran.org/hapi/ (accessed on 23 September 2024)

When computing the transmittance at a specific temperature and pressure, we search the database for the absorption coefficients of twelve gases at the two temperatures closest to the desired conditions. We then interpolate these values to obtain the absorption coefficients of the gases under the required conditions and calculate their transmittance. Finally, we multiply these transmittance values to obtain the total transmittance for the mixture. For example, to determine the transmittance of CO_2 at 273.15 K and 1000 hPa (which is included in the vertical grid), we simply find the absorption coefficients of CO_2 at 273 K and 273.5 K under 1000 hPa and perform interpolation. We then calculate the transmittance based on the newly obtained absorption coefficients. According to the Beer–Bouguer–Lambert law, the transmittance τ is defined as follows:

$$\tau_{\nu,\theta} = \exp(-k_\nu \times \mu_\theta) = \exp\left(-k_\nu \times \mathrm{Na} \times P \times L \times C \times 10^{-8}/(t \times \mathrm{R} \times \cos\theta)\right) \tag{10}$$

where k is the absorption coefficient, $1/(\text{molec}\cdot\text{cm}^{-1})$; Na = $6.022141291 \times 10^{23}$, represents Avogadro's constant; P (hPa) represents the air pressure; L (m) represents the height of gas level; C (ppmv) represents the gas concentration; t (K) represents the air temperature; and R is the ideal gas constant, R = 8.3145.

2.4.4. Simulated Spectra Correction

Due to the mismatch in resolution and wavenumber step between the simulated radiation spectrum calculated by the radiative transfer model and the observed spectrum, it is necessary to reconstruct the simulated spectrum. For example, with an observed spectrum resolution of 1 cm^{-1}, the first step is to interpolate the simulated spectrum using a specific wavenumber grid. In this study, the simulated spectrum has a step size of 0.1 cm^{-1}. After interpolation, the step size becomes 0.12055 cm^{-1}, corresponding to a spectral resolution of 0.25 cm^{-1}. Next, the reconstructed spectrum undergoes an Inverse Fourier Transform (IFT) to obtain an interferogram. Based on the resolution of the observed spectrum, a Hanning window function is applied to truncate the interferogram. Finally, a Fourier Transform (FT) is performed on the truncated interferogram to yield a simulated spectrum that matches the instrument's resolution of 1 cm^{-1}. The comparison of the simulated spectrum before and after truncation is shown in Figure 2.

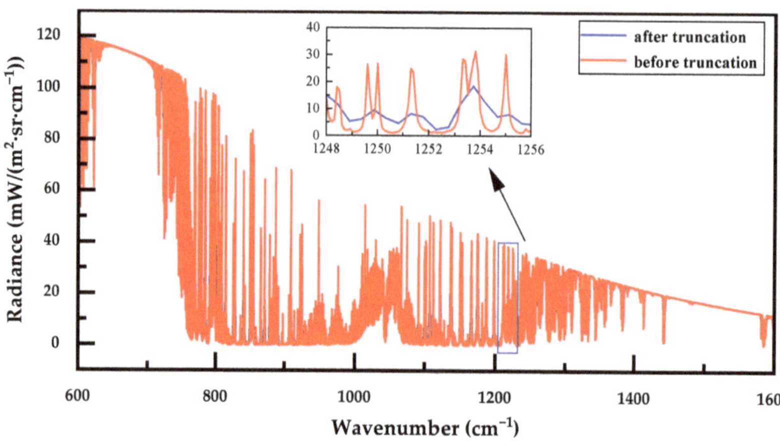

Figure 2. Comparison of the simulated spectra before and after truncation.

Figure 3 illustrates the overall research workflow of this manuscript, along with the key functions of its subcomponents.

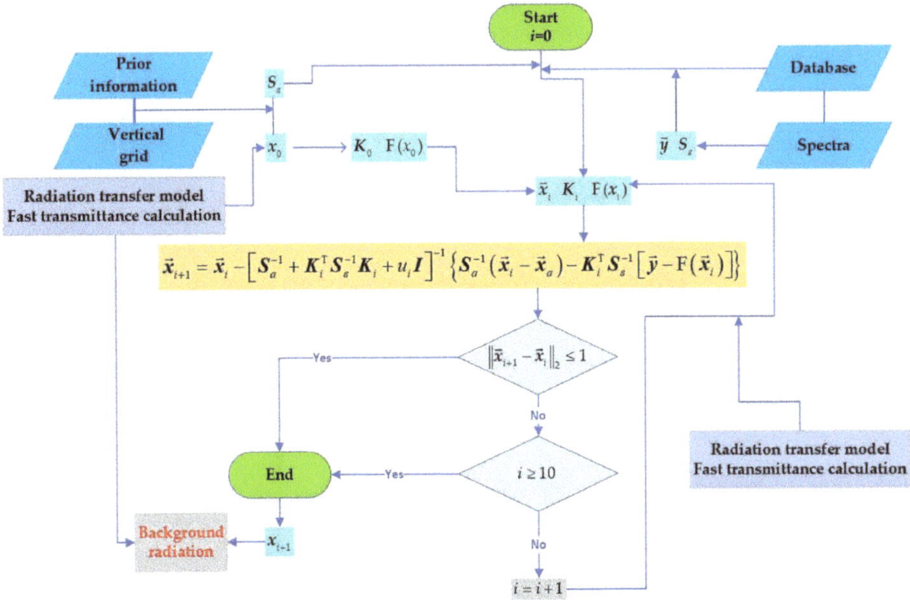

Figure 3. Schematic of the research.

3. Results and Analysis

3.1. Assessment of Rapid Transmittance Calculation Accuracy

In this paper, the temperature and humidity retrieval bands are 675–712 cm^{-1} and 1250–1350 cm^{-1}, respectively [17,18]. We utilized the HAPI and interpolation methods to calculate the mixed transmittance of 12 gases under 25 different conditions (combinations of 5 temperatures and 5 pressures). The temperature, pressure, and gas concentrations are listed in Table 2, with the optical path length set to 10 m.

Table 2. Temperature, pressure, and gas concentrations used in the assessment of rapid calculation accuracy of transmittance.

Temperature (K)	Pressure (hPa)	Gas Concentrations (ppmv)	
270.1	1000	CO_2: 400	O_2: 210000
272.3	900	H_2O: 10000	NH_3: 0.0042
274.5	800	CH_4: 1.90	NO: 0.0016
276.7	700	CO: 0.20	NO_2: 0.0060
278.9	600	N_2O: 0.33	SO_2: 0.0045
		O_3: 0.03	HNO_3: 0.0011

We compared the transmittance values obtained from these two methods in the respective bands and calculated the average deviation (AD), root mean square error (RMSE), and standard deviation of the errors (STDE) between them. Additionally, we computed the average relative deviation (ARD) for the corresponding bands (see Figures 4 and 5).

The relative error between the transmittance calculated using the interpolation method and that obtained from the HAPI program is less than 1‱. The AD, RMSE, and STDE between the two methods are also very small, indicating that the interpolation method achieves high computational accuracy. Furthermore, the computation time for calculating a single component's transmittance spectrum using the interpolation method does not exceed 1 s, whereas the HAPI program can take over ten seconds or even several tens of seconds. We recorded the time required for the two methods to calculate the mixed transmittance of 12 gases across different bands under 25 conditions, and the results are presented in Table 3. It indicates that the computational speed of the interpolation method is notably impressive.

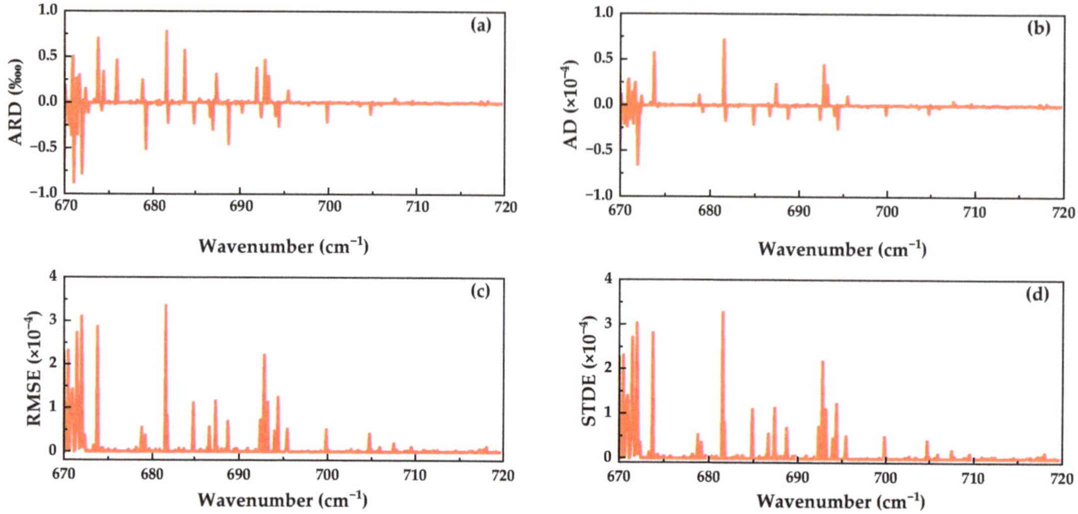

Figure 4. The ARD (**a**), AD (**b**), RMSE (**c**), and ATDE (**d**) between the transmittance values obtained from the two methods in the temperature retrieval band.

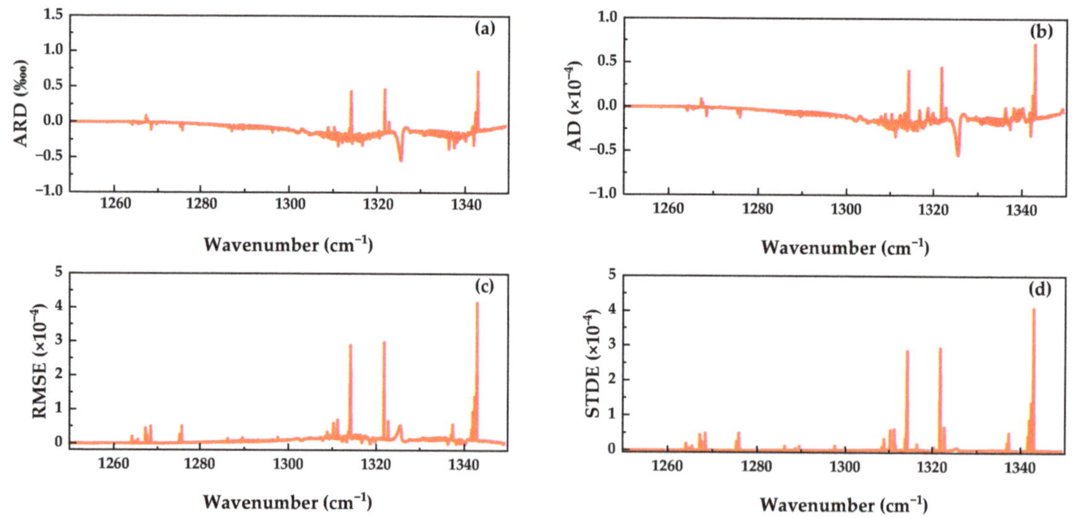

Figure 5. The ARD (**a**), AD (**b**), RMSE (**c**), and ATDE (**d**) between the transmittance values obtained from the two methods in the humidity retrieval band.

Table 3. Comparison of calculation speeds of the two methods.

	Total Time Taken (s)	
Calculation Bands (cm^{-1})	HAPI	Interpolation Method
672–715	898	37.5
1250–1350	1151	35
500–2500	3118	37.5

3.2. Evaluation of Temperature and Humidity Profile Retrieval Accuracy

Before analyzing the accuracy of the profile retrieval, we first conducted a sensitivity analysis of the proposed method. We collected 30 consecutive spectra from two different time periods using an infrared radiometer and retrieved the temperature and humidity profiles from these spectra. As shown in Figure 6, the atmospheric environmental parameters did not change significantly during a single period, and the differences between the measured spectra were small. Therefore, there were no significant differences between the retrieval results, and the standard deviation was small (see Figure 7), indicating that the method exhibits a certain degree of stability. On the other hand, the profiles retrieved from spectra measured at different times showed differences, indicating that the method is highly sensitive to changes in the environment.

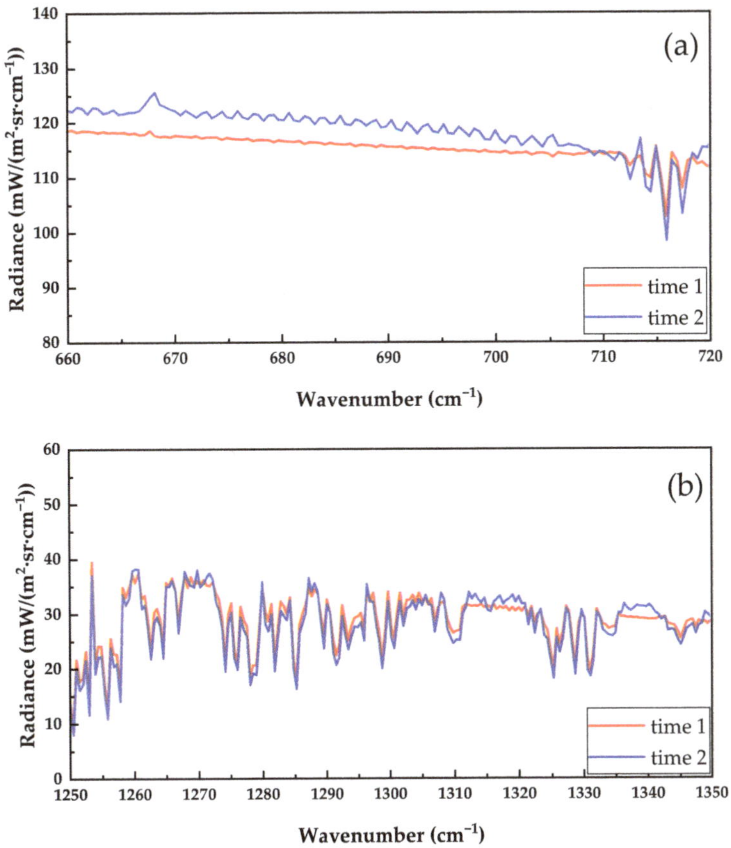

Figure 6. Average spectra measured during the two time periods: (**a**) temperature retrieval band; and (**b**) humidity retrieval band.

We obtained hourly ERA5 data from the ECMWF website for 1st, 2nd, 7th, and 8th April 2024, totaling 96 sets of data [23]. Following the method outlined in Section 2, we extracted the temperature, water vapor, and ozone profiles from these data. We then obtained average profiles for 10 additional gases from a website provided by James Hannigan (ftp://nitrogen.acom.ucar.edu/user/jamesw/IRWG/2021/WACCM.v7/Hefei.V7/, accessed on 1 September 2024) and interpolated them onto the vertical grid shown in Table 1. A simplified atmospheric model was established (see Appendix A), where temperature and water vapor were considered variable, while the other gases were treated as constant.

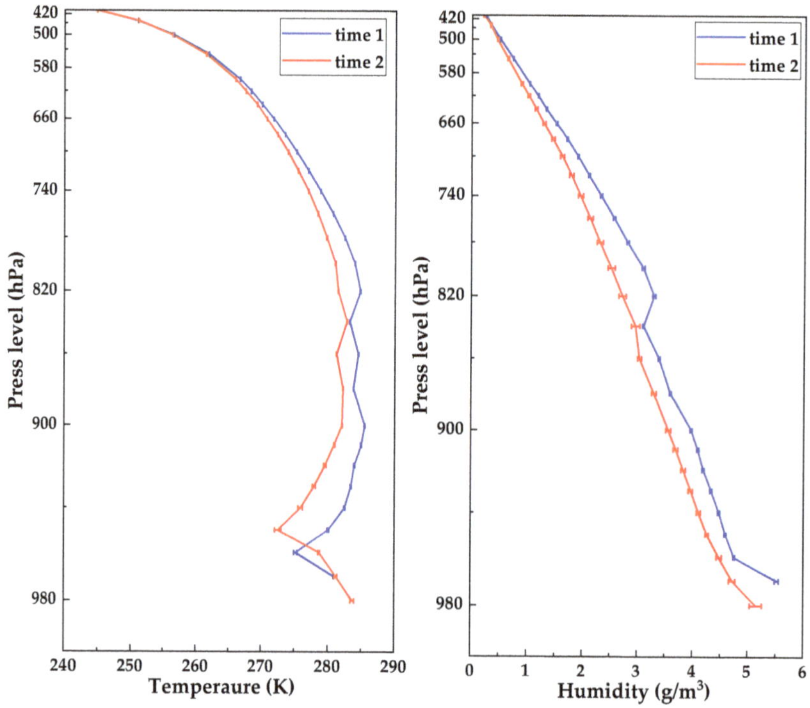

Figure 7. Comparison of the retrieval results of the spectra measured over different time periods.

The retrieval of temperature and humidity is performed sequentially. In the temperature retrieval process, the temperature profile is treated as a variable while keeping the other parameters constant. The water vapor and ozone profiles are set to the average values of the priori profiles, and the other parameters are set according to Appendix A. In the humidity retrieval process, the humidity profile is treated as a variable, while the temperature profile is set to the values retrieved in the temperature retrieval step. The remaining parameters are kept the same as in the temperature retrieval process.

Based on the aforementioned 96 sets of data, we performed simulation calculations to obtain 96 radiative spectra. These spectra were then used for fast retrieval of temperature and humidity profiles. Subsequently, we conducted an analysis and evaluation of the retrieval results. It is worth mentioning that with the use of the rapid retrieval method, the average time required to obtain a temperature profile and a humidity profile does not exceed 5 min, which fully meets the requirements for engineering applications.

Figure 8 presents the average deviation (AD), root mean square error (RMSE), and standard deviation of the errors (STDE) for the temperature and humidity profile retrieval results compared to the theoretical values. For temperature retrieval, the difference between the retrieval results and the theoretical values is relatively small near the surface but increases with altitude. Previous studies have indicated that retrieval accuracy is noticeably higher below the boundary layer than above it, which aligns with our expectations for the rapid temperature profile retrieval results.

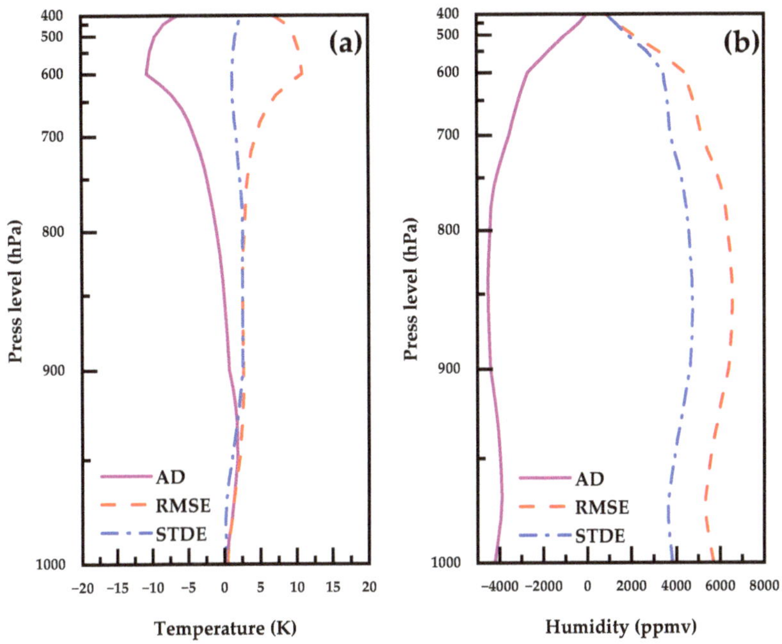

Figure 8. The AD, RMSE, and STDE of temperature (**a**) and humidity (**b**) retrievals.

In contrast, for humidity retrieval, we observed a significant difference throughout the entire altitude range. On one hand, due to the saturation of water vapor absorption, the retrieval results tend to be significantly lower than the true values when humidity is high. On the other hand, it is possible that the method used in this study itself introduces substantial errors in humidity retrieval; however, we are currently unable to pinpoint the exact source of these errors.

3.3. Evaluation of Sky Background Radiation Simulations

As mentioned earlier, the current method based on simulated atmospheric background radiation has some latency. To demonstrate the real-time advantage of the rapid retrieval method for obtaining atmospheric background radiation, we used 48 simulated spectra corresponding to the data from 7 and 8 April 2024, as hypothetical measured spectra. We considered the ECMWF results from 1 and 2 April as the input with latency, while the retrieval results from the 7 and 8 April spectra served as the input with real-time information. By comparing the differences between the sky background radiation obtained from these two parameter inputs and the hypothetical measured spectra, we can highlight the advantage of real-time retrieval of sky background radiation using the rapid retrieval method.

We calculated the deviations, RMSE, and STDE for the two sets of 48 sky background radiation spectra compared to the hypothetical measured spectra. From Figures 9 and 10, it is evident that the RMSE and STDE of the sky background radiation calculated using the rapid retrieval method with real-time input profiles are significantly lower than those obtained using the latest profiles provided by ECMWF ERA5. Figure 11 illustrates the sky background radiation calculated by both methods at a specific time. It is visually apparent that the background radiation obtained using the rapid retrieval method is more reasonable and closer to the hypothetical measured spectra.

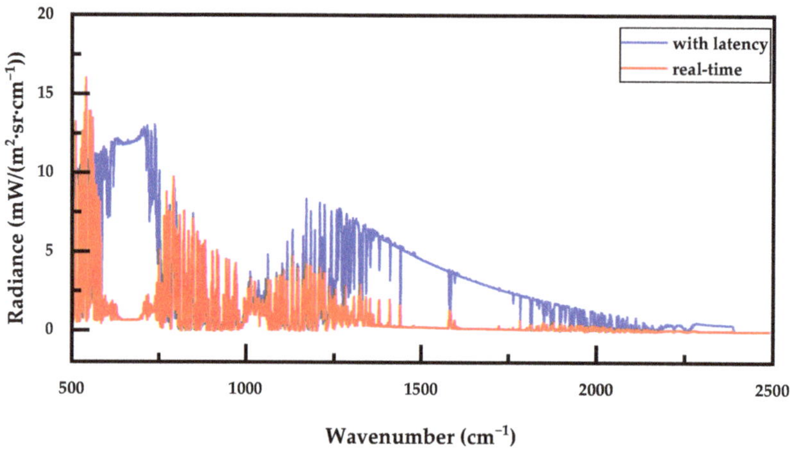

Figure 9. The RMSE between the measured spectra and the two types of background radiation spectra. Blue line: background radiation calculated with the hysteresis profile provided by ERA5 as the input parameter; red line: background radiation calculated with the real-time profile obtained by the fast retrieval method as the input parameter.

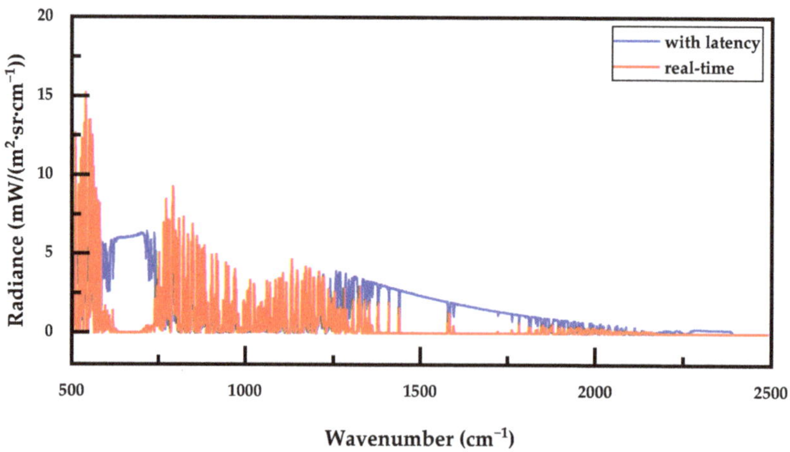

Figure 10. As in Figure 9 but for STDE.

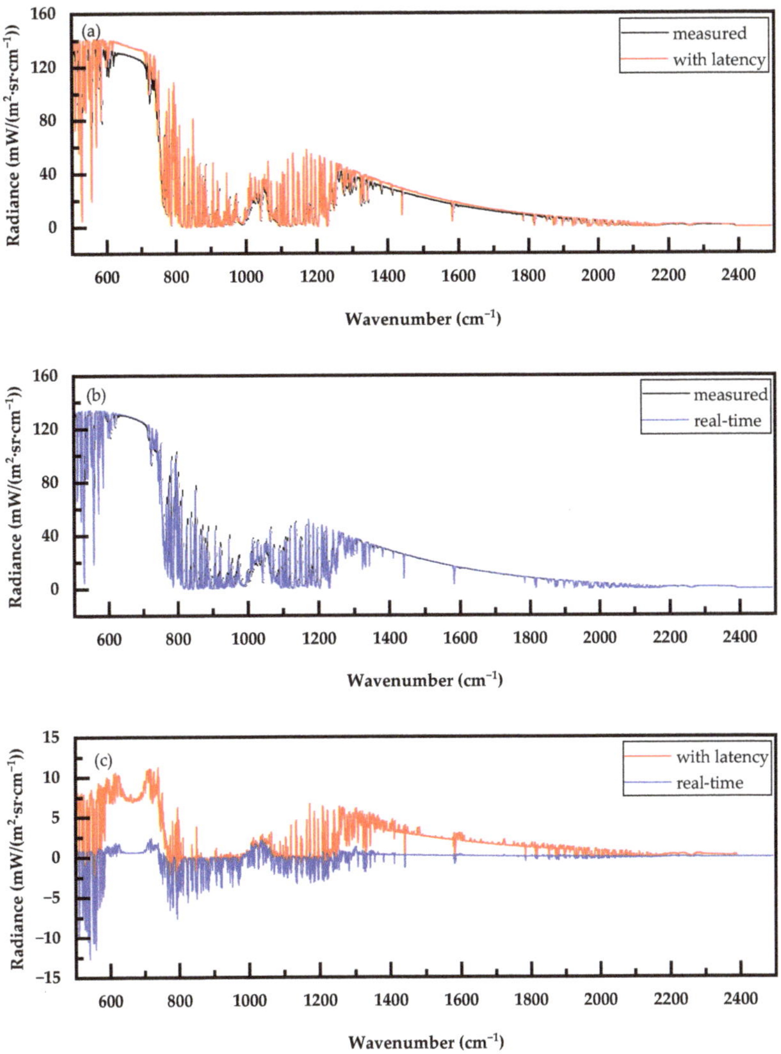

Figure 11. The difference between the measured spectra and the two types of background radiation spectra at 0:00 UTC on 7th April 2024: (**a**) background radiation with latency; (**b**) background radiation in real-time; (**c**) bias of two types of background radiation spectra with measured spectra.

4. Conclusions and Discussion

At present, researchers mainly acquire sky background radiation for application in infrared remote sensing in two ways. The first method is to periodically collect and update the background spectrum in the monitoring application process, which is simple to operate but will interrupt the monitoring process. Therefore, the second method is derived to simulate the background radiation based on the atmospheric parameter profile and transmission model. This method can update the background spectrum in the monitoring process, but there is a lag problem in obtaining the atmospheric parameter profile. In order to solve this problem, based on the theoretical framework of the existing infrared remote sensing technology, we simplified the retrieval method of temperature and humidity profiles, focused on the transmission calculation and transmission model, and established a fast and real-time method to obtain the sky background radiation.

The fast transmittance calculation method utilizing line-by-line integration and interpolation principles significantly improves computational speed while maintaining high accuracy compared to the HAPI program. Additionally, the simplified transmission model, which incorporates 12 gas components, is sufficient for simulating background radiation. With this new method, we can rapidly extract temperature and humidity profile information corresponding to the measured spectra and obtain the associated background spectra. Compared to extracting the latest profiles from datasets, the new method shows smaller differences between the calculated background radiation and measured spectra, making it more suitable for calculating sky background radiation.

Furthermore, we compare the differences between the retrieval results of the profiles and the true values. The results show a certain level of accuracy in temperature retrieval of the lower atmosphere, but the humidity retrieval results are not ideal. In addition, the current method can only be applied to clear sky conditions, and cannot be effectively processed for clouds. Considering that this temperature and humidity fast retrieval method does not require high hardware and software requirements, we plan to conduct further research on the accuracy of humidity and cloud retrieval in the follow-up work. We hope that this fast retrieval method can be widely applied in temperature and humidity profile observation applications and scientific research.

Author Contributions: Conceptualization, S.S. and L.X.; Formal analysis, S.S. and L.X.; Funding acquisition, J.L. and L.X.; Methodology, S.S.; Resources, J.L. and L.X.; Software, S.S., Y.W. and Y.S.; Validation, Y.D.; Writing—original draft, S.S.; Writing—review and editing, S.S. and L.X. All authors have read and agreed to the published version of the manuscript.

Funding: This project is supported by the key R&D program of Anhui Province, grant number 2022m07020009, and the National Natural Science Foundation of China, grant number 52027804.

Institutional Review Board Statement: Not applicable.

Informed Consent Statement: Not applicable.

Data Availability Statement: The original contributions presented in the study are included in the article, further inquiries can be directed to the corresponding author.

Acknowledgments: The authors would like to thank the European Centre for Medium-Range Weather Forecasts and the Infrared Working Group for sharing the basic research data. We also would like to thank anonymous reviewers for their insightful comments and constructive suggestions.

Conflicts of Interest: The authors declare no conflicts of interest.

Appendix A

Table A1. The pressure (P, hPa), height (H, m), and concentration (ppmv) of gases in each layer of the model atmosphere.

P	H	CO_2	CH_4	CO ($\times10^{-1}$)	N_2O ($\times10^{-1}$)	HNO_3 ($\times10^{-3}$)	NH_3 ($\times10^{-3}$)	NO ($\times10^{-3}$)	NO_2 ($\times10^{-3}$)	O_2 ($\times10^{5}$)	SO_2 ($\times10^{-3}$)
1000	50	402.78	1.94	1.96	3.25	1.08	4.23	1.59	7.55	21.11	4.51
990	150	402.74	1.93	1.85	3.25	1.19	3.80	1.38	6.59	21.11	4.00
980	300	402.70	1.93	1.74	3.25	1.08	3.38	1.17	5.63	21.11	3.49
970	420	402.67	1.93	1.63	3.25	1.19	2.95	0.96	4.67	21.11	2.98
960	530	402.63	1.92	1.52	3.25	1.30	2.52	0.75	3.71	21.12	2.48
950	600	402.59	1.92	1.41	3.25	1.40	2.09	0.55	2.75	21.12	1.97
940	710	402.55	1.92	1.30	3.25	1.51	1.66	0.34	1.79	21.12	1.46
930	810	402.51	1.91	1.19	3.25	1.62	1.23	0.13	0.83	21.12	0.95
920	900	402.50	1.91	1.17	3.25	1.54	1.14	0.11	0.73	21.12	0.87
910	1000	402.49	1.91	1.14	3.25	1.46	1.05	0.09	0.62	21.12	0.79
900	1100	402.47	1.91	1.11	3.25	1.38	0.95	0.07	0.51	21.12	0.71
880	1280	402.45	1.91	1.06	3.25	1.23	0.77	0.04	0.29	21.13	0.54

Table A1. *Cont.*

P	H	CO_2	CH_4	CO ($\times 10^{-1}$)	N_2O ($\times 10^{-1}$)	HNO_3 ($\times 10^{-3}$)	NH_3 ($\times 10^{-3}$)	NO ($\times 10^{-3}$)	NO_2 ($\times 10^{-3}$)	O_2 ($\times 10^{5}$)	SO_2 ($\times 10^{-3}$)
860	1460	402.44	1.90	1.02	3.25	1.09	0.67	0.03	0.23	21.13	0.47
840	1660	402.43	1.90	0.99	3.25	0.95	0.57	0.02	0.17	21.13	0.39
820	1860	402.42	1.90	0.96	3.25	0.83	0.49	0.02	0.12	21.14	0.33
800	2080	402.42	1.90	0.94	3.25	0.73	0.43	0.01	0.11	21.14	0.29
780	2280	402.42	1.89	0.91	3.25	0.63	0.36	0.01	0.09	21.14	0.25
760	2500	402.44	1.89	0.90	3.25	0.56	0.31	0.01	0.07	21.15	0.22
740	2720	402.46	1.89	0.88	3.25	0.49	0.26	0.01	0.07	21.15	0.19
720	2940	402.49	1.89	0.86	3.25	0.42	0.22	0.01	0.06	21.16	0.16
700	3180	402.54	1.89	0.85	3.25	0.38	0.18	0.01	0.05	21.16	0.14
680	3420	402.59	1.89	0.84	3.26	0.33	0.15	0.01	0.05	21.17	0.12
660	3650	402.64	1.89	0.82	3.26	0.29	0.12	0.01	0.04	21.17	0.10
640	3900	402.69	1.88	0.81	3.26	0.27	0.10	0.01	0.04	21.17	0.09
620	4150	402.74	1.88	0.80	3.26	0.24	0.08	0.01	0.04	21.18	0.07
600	4420	402.79	1.88	0.80	3.26	0.23	0.07	0.01	0.04	21.18	0.06
550	5120	402.92	1.88	0.78	3.26	0.20	0.04	0.01	0.04	21.19	0.04
500	5860	403.04	1.88	0.77	3.26	0.19	0.03	0.01	0.04	21.20	0.03
450	6670	403.09	1.88	0.77	3.26	0.23	0.02	0.01	0.05	21.20	0.02
400	7550	403.07	1.87	0.77	3.26	0.27	0.02	0.02	0.06	21.21	0.02
350	8540	402.96	1.87	0.76	3.26	0.32	0.02	0.03	0.07	21.21	0.02
300	9630	402.74	1.87	0.75	3.26	0.36	0.02	0.04	0.09	21.21	0.02
250	10,880	402.37	1.86	0.71	3.25	0.45	0.01	0.06	0.11	21.21	0.02
200	12,380	401.78	1.85	0.62	3.23	0.63	0.01	0.08	0.12	21.22	0.02
175	13,250	401.45	1.83	0.55	3.22	0.75	0.01	0.08	0.13	21.22	0.02
150	14,230	401.14	1.82	0.47	3.19	0.89	0.01	0.09	0.13	21.23	0.02
125	15,350	400.93	1.81	0.39	3.17	1.04	0.00	0.10	0.16	21.23	0.02
100	16,720	400.64	1.79	0.31	3.12	1.35	0.00	0.13	0.21	21.23	0.03
75	18,410	399.67	1.73	0.21	2.99	2.25	0.00	0.17	0.33	21.23	0.07
50	20,850	396.89	1.55	0.13	2.55	5.01	0.00	0.27	0.69	21.23	0.09
30	24,090	394.46	1.32	0.15	1.88	7.71	0.00	0.62	1.91	21.23	0.02
25	25,250	394.04	1.25	0.16	1.67	7.85	0.00	0.85	2.65	21.22	0.02
20	26,720	393.69	1.17	0.17	1.42	7.35	0.00	1.27	3.85	21.22	0.02
15	28,600	393.45	1.07	0.19	1.12	5.91	0.00	2.04	5.70	21.22	0.02
10	31,320	393.32	0.96	0.22	0.79	3.55	0.00	3.37	8.10	21.22	0.02
7	33,800	393.31	0.88	0.26	0.57	1.92	0.00	4.59	9.39	21.22	0.01
5	36,160	393.33	0.81	0.31	0.42	0.96	0.00	5.67	9.72	21.22	0.01
4	37,840	393.34	0.76	0.34	0.35	0.59	0.00	6.38	9.51	21.22	0.01
3	40,000	393.35	0.70	0.38	0.27	0.29	0.00	7.14	9.00	21.22	0.02
2	43,100	393.33	0.60	0.44	0.18	0.09	0.00	7.72	7.92	21.22	0.03

References

1. Capodaglio, A.G. Developments and Issues in Renewable Ecofuels and Feedstocks. *Energies* **2024**, *17*, 3560. [CrossRef]
2. Wu, M.; Liu, W.; Ma, Z.; Qin, T.; Chen, Z.; Zhang, Y.; Cao, N.; Xie, X.; Chi, S.; Xu, J.; et al. Global Trends in the Research and Development of Petrochemical Waste Gas from 1981 to 2022. *Sustainability* **2024**, *16*, 5972. [CrossRef]
3. Zhao, Q.; Nie, X.X.; Luo, D.; Wang, J.; Li, Q.R.; Chen, W. An Effective Method for Gas-Leak Area Detection and Gas Identification with Mid-Infrared Image. *Photonics* **2022**, *9*, 992. [CrossRef]
4. Rodriguez-Conejo, M.A.; Melendez, J.; Barber, R.; Garrido, S. Design of an Infrared Imaging System for Robotic Inspection of Gas Leaks in Industrial Environments. *Int. J. Adv. Robot. Syst.* **2015**, *12*, 23. [CrossRef]
5. Hu, Y.Y.; Xu, L.; Xu, H.Y.; Shen, X.C.; Deng, Y.S.; Xu, H.Y.; Liu, J.G.; Liu, W.Q. Three-dimensional reconstruction of a leaking gas cloud based on two scanning FTIR remote-sensing imaging systems. *Opt. Express* **2022**, *30*, 25581–25596. [CrossRef] [PubMed]
6. Hu, Y.Y.; Xu, L.; Shen, X.C.; Jin, L.; Xu, H.Y.; Deng, Y.S.; Liu, J.G.; Liu, W.Q. Reconstruction of a leaking gas cloud from a passive FTIR scanning remote-sensing imaging system. *Appl. Opt.* **2021**, *60*, 9396–9403. [CrossRef] [PubMed]
7. Donato, P.; Barres, O.; Sausse, J.; Taquet, N. Advances in 3-D infrared remote sensing gas monitoring. Application to an urban atmospheric environment. *Remote Sens. Environ.* **2016**, *175*, 301–309. [CrossRef]
8. Kastek, M.; Piatkowski, T.; Trzaskawka, P. Infrared Imaging Fourier Transform Spectrometer as the Stand-Off Gas Detection System. *Metrol. Meas. Syst.* **2011**, *18*, 607–620. [CrossRef]

9. Stremme, W.; Krueger, A.; Harig, R.; Grutter, M. Volcanic SO_2 and SiF_4 visualization using 2-D thermal emission spectroscopy–Part 1: Slant-columns and their ratios. *Atmos. Meas. Tech.* **2012**, *5*, 275–288. [CrossRef]
10. Kang, R.Y.; Liatsis, P.; Kyritsis, D.C. Emission Quantification via Passive Infrared Optical Gas Imaging: A Review. *Energies* **2022**, *15*, 3304. [CrossRef]
11. Wu, S.P.; Zhong, X.; Qu, Z.; Wang, Y.H.; Li, L.; Zeng, C.L. Infrared Gas Detection and Concentration Inversion Based on Dual-Temperature Background Points. *Photonics* **2023**, *10*, 490. [CrossRef]
12. Hu, Y.Y.; Xu, L.; Shen, X.C.; Shu, S.Q.; Xu, H.Y.; Deng, Y.S.; Xu, H.Y.; Liu, J.G.; Liu, W.Q. Inversion method of target gas cloud transmittance based on atmospheric profile synthesis background. *Acta Phys. Sin.* **2023**, *72*, 033201. [CrossRef]
13. Knuteson, R.; Revercomb, H.; Best, F.; Ciganovich, N.; Dedecker, R.; Dirkx, T.; Ellington, S.; Feltz, W.; Garcia, R.; Howell, H.; et al. Atmospheric Emitted Radiance Interferometer. Part I: Instrument Design. *J. Atmos. Ocean. Technol.* **2004**, *21*, 1763–1776. [CrossRef]
14. Knuteson, R.; Revercomb, H.; Best, F.; Ciganovich, N.; Dedecker, R.; Dirkx, T.; Ellington, S.; Feltz, W.; Garcia, R.; Howell, H.; et al. Atmospheric Emitted Radiance Interferometer. Part II: Instrument Performance. *J. Atmos. Ocean. Technol.* **2004**, *21*, 1777–1789. [CrossRef]
15. Zhao, H.J.; Ma, X.H.; Jia, G.R.; Mi, Z.Y.; Ji, H.L. Synergistic Retrieval of Temperature and Humidity Profiles from Space-Based and Ground-Based Infrared Sounders Using an Optimal Estimation Method. *Remote Sens.* **2022**, *14*, 5256. [CrossRef]
16. Löhnert, U.; Turner, D.D.; Crewell, S. Ground-Based Temperature and Humidity Profiling Using Spectral Infrared and Microwave Observations. Part I: Simulated Retrieval Performance in Clear-Sky Conditions. *J. Appl. Meteorol. Climatol.* **2009**, *48*, 1017–1032. [CrossRef]
17. Smith, W.L.; Feltz, W.E.; Knuteson, R.O.; Revercomb, H.E.; Woolf, H.M.; Howell, H.B. The retrieval of planetary boundary layer structure using ground-based infrared spectral radiance measurements. *J. Atmos. Ocean. Technol.* **1999**, *16*, 323–333. [CrossRef]
18. Feltz, W.F.; Smith, W.L.; Knuteson, R.O.; Revercomb, H.E.; Woolf, H.M.; Howell, H.B. Meteorological applications of temperature and water vapor retrievals from the ground-based Atmospheric Emitted Radiance Interferometer (AERI). *J. Appl. Meteorol.* **1998**, *37*, 857–875. [CrossRef]
19. Rodgers, C. Optimal Methods for Non-linear Inverse Problems. In *Inverse Methods for Atmospheric Sounding: Theory and Practice*; Taylor, F.W., Ed.; World Scientific: London, UK, 2000; Volume 2, pp. 81–99.
20. Zhou, X.H.; Liang, Y.; Wang, X.M.; Zhou, L.C. Comparison of saturation vapor pressure formulas. *J. Liaoning Tech. Univ.* **2007**, *26*, 331–333.
21. Liou, K. Thermal Infrared Radiation Transfer in the Atmosphere. In *An Introduction to Atmospheric Radiation*, 2nd ed.; Dmowska, R., Holton, J.R., Rossby, H.T., Eds.; Academic Press: San Diego, CA, USA, 2002; Volume 84, pp. 122–127.
22. Kochanov, R.V.; Gordon, I.; Rothman, L.; Wcisło, P.; Hill, C.; Wilzewski, J. HITRAN Application Programming Interface (HAPI): A comprehensive approach to working with spectroscopic data. *J. Quant. Spectrosc. Radiat. Transf.* **2016**, *177*, 15–30. [CrossRef]
23. Hersbach, H.; Bell, B.; Berrisford, P.; Biavati, G.; Horányi, A.; Muñoz Sabater, J.; Nicolas, J.; Peubey, C.; Radu, R.; Rozum, I.; et al. ERA5 Hourly Data on Pressure Levels from 1940 to Present. Copernicus Climate Change Service (C3S) Climate Data Store (CDS). Available online: https://cds.climate.copernicus.eu/cdsapp#!/dataset/10.24381/cds.bd0915c6?tab=overview (accessed on 23 June 2024).

photonics

MDPI

Article

Performance Evaluation and Error Tracing of Rotary Rayleigh Doppler Wind LiDAR

Jianfeng Chen [1,2,3], Chenbo Xie [2,3,*], Jie Ji [1,2,3], Leyong Li [1,2,3], Bangxin Wang [2,3], Kunming Xing [2,3] and Ming Zhao [4,*]

[1] Science Island Branch, University of Science and Technology of China, Hefei 230026, China; jianfengchen1212@mail.ustc.edu.cn (J.C.); jijie3496@mail.ustc.edu.cn (J.J.); lly9977@mail.ustc.edu.cn (L.L.)
[2] Key Laboratory of Atmospheric Optics, Anhui Institute of Optics and Fine Mechanics, Chinese Academy of Sciences, Hefei 230031, China; bxwang@aiofm.ac.cn (B.W.); kunmingx@mail.ustc.edu.cn (K.X.)
[3] Advanced Laser Technology Laboratory of Anhui Province, Hefei 230037, China
[4] School of Electronic Engineering, Huainan Normal University, Huainan 232038, China
* Correspondence: cbxie@aiofm.ac.cn (C.X.); zhaom@hnnu.edu.cn (M.Z.)

Abstract: In the study of atmospheric wind fields from the upper troposphere to the stratosphere (10 km to 50 km), direct detection wind LiDAR is considered a promising method that offers high-precision atmospheric wind field data. In 2020, Xie et al. of the Anhui Institute of Optics and Fine Mechanics, Chinese Academy of Sciences, developed an innovative rotating Rayleigh Doppler wind LiDAR (RRDWL). The system aims to achieve single-LiDAR detection of atmospheric wind fields by rotating the entire device cabin. In 2022, the feasibility of the system was successfully validated in laboratory conditions, and field deployment was completed. Due to the structural differences between this system and traditional direct-detection wind LiDAR, performance tests were conducted to evaluate its continuous detection capability in outdoor environments. Subsequently, based on the test results and error analysis, further analysis was carried out to identify the main factors affecting the system's detection performance. Finally, the error analysis and traceability of the detection results were conducted, and corresponding measures were discussed to provide a theoretical foundation for optimizing the performance of RRDWL.

Keywords: rotary Rayleigh doppler wind LiDAR; performance evaluation; error tracing

1. Introduction

Wind is a physical phenomenon on Earth that involves the horizontal movement of air, describing the motion of air currents. In scientific terms, wind is described as the moving component of air, encompassing both wind speed and wind direction. As a significant concept in meteorology, a wind field refers to the distribution of wind speed and wind direction at various positions in the atmosphere during a specific time and within a particular spatial range. The study of atmospheric wind fields has its roots in long-term observations and explorations of weather, climate, and atmospheric motions. Accurate understanding and prediction of weather conditions are crucial for human life, agriculture, transportation, aviation, and other aspects [1–5]. Currently, there is a widespread recognition of the significance of measuring the tropospheric wind field in improving the accuracy of numerical weather prediction and wind shear warnings. The stratosphere, located above the troposphere and extending from approximately 10 km to 50 km in height, plays a vital role in the climate system and atmospheric stability. Therefore, the importance of measuring the stratospheric wind field cannot be overlooked.

However, atmospheric wind fields exhibit complex non-uniformity and spatiotemporal variability. Wind speed and wind direction can vary significantly at different altitudes, geographical locations, and over time. This non-uniformity poses challenges in accurately measuring and depicting the complete wind field. Weather conditions, terrain, and environmental factors also play important roles in limiting wind field detection. For instance, strong winds, precipitation, cloud cover, and complex terrain can all result in incomplete and inaccurate observational data. Additionally, the detection of wind fields requires the use of multiple measurement devices and techniques; however, different equipment and methods have limitations in terms of measurement range, resolution, accuracy, and reliability [6–10]. In summary, the characterization of wind fields is a complex task due to these factors. Light Detection and Ranging (LiDAR) is an advanced technology that utilizes laser beams and scattered light for precise measurement and observation. In comparison to traditional meteorological sounding methods, LiDAR offers the advantage of non-contact observations, allowing for a wider range of measurements, including higher altitude levels and complex terrain areas [11–13].

Direct-detection wind LiDAR is a detection method that utilizes a frequency discriminator, also known as an optical mixer, to convert the Doppler frequency shift of aerosols or molecules caused by wind in the atmosphere into changes in light energy [14–16]. Based on the type of scatterers, direct-detection wind LiDAR can be categorized into three types: Mie wind LiDAR, Rayleigh wind LiDAR, and resonance fluorescence wind LiDAR. Compared to aerosol scattering, Rayleigh scattering is more prevalent and relatively stronger in the upper atmosphere. Despite the limitations in sensitivity due to the width of the molecular backscatter spectrum, Rayleigh scattering wind LiDAR can effectively detect the wind field in the global free troposphere and upper atmosphere [17–19]. Traditionally, direct-detection wind LiDAR has employed two methods to measure different line of sight (LOS) wind speeds. The first method involves mounting a scanning head above the LiDAR transceiver, which is rotated to achieve a cone scan. However, this approach becomes challenging for larger systems when the aperture of the receiving telescope exceeds half a meter. The required size, thickness, and weight of the scanning mirror increase rapidly, making it difficult to handle and more expensive. As a result, this method is primarily suitable for small and medium-sized systems [20,21]. In order to address the challenges of measuring high-altitude wind fields, researchers have explored an alternative approach. They have employed two large aperture LiDAR subsystems placed orthogonal to each other to measure orthogonal LOS wind speeds [22–24]. However, this approach introduces certain issues. Using two subsystems necessitates the use of two lasers, multiple large-aperture telescopes, and two separate detection and acquisition systems. This increases the costs of development and maintenance, as well as the complexity of calibrating and fine-tuning these precise instruments.

Based on the aforementioned background, Xie et al. from the Anhui Institute of Optics and Fine Mechanics, Chinese Academy of Sciences, have developed a rotary Rayleigh Doppler wind LiDAR (RRDWL). The system innovatively uses a rotating platform structure to detect horizontal wind fields through the overall rotation of the equipment. This design not only makes it possible for a single direct wind measurement LiDAR system to detect the atmospheric wind field but also effectively reduces the development cost of the equipment. In 2020, the system verified the feasibility of detecting atmospheric wind fields in a laboratory environment. Subsequently, the outfield construction was completed in 2022 [25,26]. However, due to the structural differences between this system and traditional direct wind LiDAR, it is necessary to evaluate its performance in the actual detection of the atmospheric wind field. In addition, due to the heterogeneity of the actual atmosphere and potential systematic errors in the system, complete consistency between detection results and simulation results cannot be achieved. Therefore, it is imperative to analyze and trace any errors in the detection results while discussing corresponding measures to establish a theoretical foundation for optimizing RRDWL's performance.

2. RRDWL Wind Measurement Principle

2.1. Rayleigh Scattering LiDAR Equation

Since the advent of using lasers as a light source for atmospheric detection, researchers have observed the interaction between lasers and atmospheric substances, including scattering and absorption phenomena. These interactions can be categorized into different physical mechanisms such as scattering (Mie, Raman, Rayleigh), resonance fluorescence, and more. In the middle atmosphere, Rayleigh scattering is a prominent form of scattering due to the relatively low aerosol content above the boundary layer and the prevalence of small diameter molecules (less than one-tenth of the laser wavelength) in the atmosphere. Rayleigh scattering signals are stronger compared to Mie scattering signals. Furthermore, the shorter the wavelength of the laser used, the stronger the Rayleigh scattering and the greater the change in scattering cross-section. As a Rayleigh scattering LiDAR primarily used for middle atmosphere detection, the backscattered signal consists of two main components. The first is the Rayleigh scattering signal, which is generated when the laser interacts with air molecules (nitrogen and oxygen) in the atmosphere. The second is the meter scattering signal, which is generated when the laser interacts with aerosol particles suspended in the lower atmosphere. Although both types of scattering are considered elastic scattering and do not involve frequency drift, there are distinct differences in their scattering processes. During the actual detection and data inversion process, it is necessary to separate the pure Rayleigh scattering signal from the atmospheric echo signal by subtracting the aerosol Mie scattering signal using atmospheric models or empirical formulas. This is because the presence of aerosol scattering interferes with the received signal, and we aim to isolate the Rayleigh scattering signals associated with atmospheric molecules in order to accurately estimate atmospheric parameters.

The RRDWL primarily focuses on the detection of the wind field in the middle atmosphere, which includes the high troposphere and stratosphere. Therefore, it falls under the category of "Rayleigh scattering Doppler wind measurement". Specifically, for the wind field in the middle atmosphere, direct wind LiDAR is the only method available for continuous observation. Therefore, the Rayleigh scattering LiDAR equation can be described using the following expression:

$$N_R(\lambda,z) = \left[\frac{P_L(\lambda)\Delta t}{hc/\lambda}\right][\sigma_R(\pi,\lambda)n_R(z)\Delta z]\left(\frac{A}{z^2}\right)T_{a,m}^2(\lambda,z)[\eta(\lambda)G(z)] + N_B. \quad (1)$$

where P_L represents the transmission power of the LiDAR system, Δt represents the integration time, σ_R represents the differential backscattering cross-section, n_R represents the number density of fully mixed air molecules, A/z^2 represents the probability of the photon being received by the telescope, $T_{a,m}^2$ represents the atmospheric transmittance of the aerosol and molecules, η represents the efficiency of the receiving optical system, G represents the geometric factor, and N_B represents the background noise.

2.2. Double Edge Technique and FPI Frequency Identification

In general, most direct-detection wind LiDAR systems employ solid-state lasers that emit laser beams with narrow linewidth Gaussian spectral lines, and it is crucial to use an optical filter with a similarly steep spectral line. This enables the distinction of changes in spectral line transmittance, which can then be converted into Doppler shifts for wind measurement purposes. Currently, Fabry–Perot interferometers (FPI) are widely used for detecting atmospheric wind fields at middle and high altitudes. The rotary Rayleigh Doppler wind LiDAR (RRDWL) utilizes a double edge technology. This involves differentially processing the laser signal emitted by the laser as a reference signal with the atmospheric backscattered light signal received by the telescope. Figure 1 illustrates this concept.

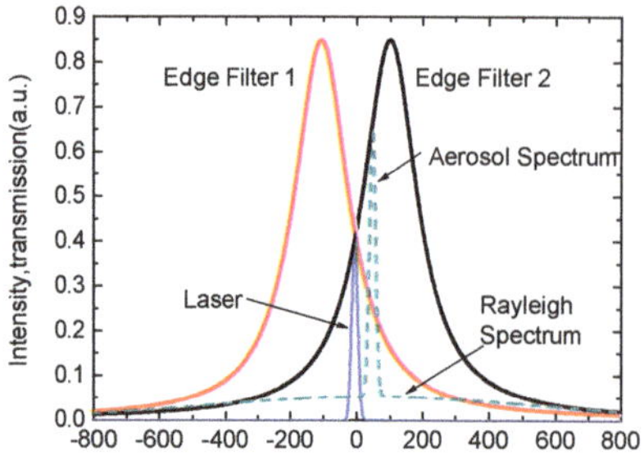

Figure 1. Double edge technology schematic diagram.

The system utilizes a tunable FPI with an adjustable optical path between two plates. The FPI is divided into three channels, and the total aperture of the light-passing part is 80 mm. Two of the channels are dedicated to measuring the atmospheric echo signal, while the third channel is used for frequency locking, ensuring that the transmittance line of the FPI is aligned with the laser frequency. However, due to slight differences in film thickness, the optical path difference between the plates in each channel will vary slightly. As a result, the transmittance spectral lines of the three channels become independent of each other, allowing for a separate analysis and processing of their respective signals. In Figure 2, the principle of frequency discrimination using the FPI is illustrated. The blue and cyan lines represent the two edge channels, while the black lines represent the locked channel. The signal intensity in the edge channels is determined by the convolution of the atmospheric echo and the spectral response of the edge channels. The Doppler shift alters the relative position of the atmospheric backscattered spectrum, causing the signals in the edge channels to rise or fall accordingly. The green line in the middle represents the initial frequency of the emitted laser. The two blue areas represent the signal spectra after passing through the two edge channels, with the area representing the signal strength. In the absence of wind, when there is no Doppler shift, the signal spectra in the two edge channels are symmetric with each other. However, in the presence of wind, the atmospheric backscattered light signal experiences a Doppler frequency shift, resulting in changes in the signals of the two edge channels. By comparing the intensity changes between the emitted laser signal and the atmospheric backscattered light signal, the Doppler shift information can be derived. This information is then used to determine the radial wind speed at the corresponding height. Essentially, by using the FPI as a frequency discriminator, the radial wind speed can be obtained simply by analyzing the intensity changes in the two signals.

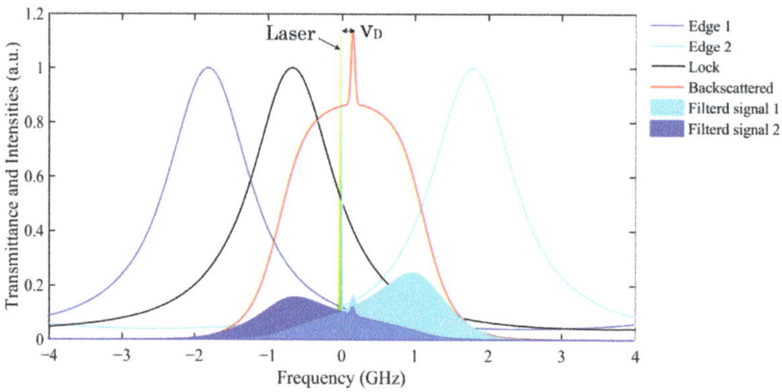

Figure 2. Schematic diagram of FPI frequency discrimination principle.

2.3. Wind Velocity Inversion and Synthesis

During actual wind field detection, the atmospheric echo signal passing through the FPI edge channel can be obtained by convolving the transmittance spectral lines of the two edge channels with the backscattered signal spectral lines of atmospheric molecules.

$$T_{Ri}(\mathrm{v}) = h(\mathrm{v}) \otimes f_L(\mathrm{v}) \otimes f_{Ray}(\mathrm{v}), \tag{2}$$

where $f_L(v)$ represents the spectral line of the emitted laser, $h(v)$ represents the actual transmittance function of FPI, $f_{Ray}(v)$ represents the Rayleigh backscattering spectrum after the interaction of atmospheric molecules with the laser, $i = 1,2$ represents the two edge channels of FPI, and "$\otimes$" represents the convolution. According to the sum of the transmittance after the convolution of the two channels, the response curve of the system to the frequency shift can be calculated:

$$R_V(\mathrm{v}_d, T) = T_{R1}(\mathrm{v}_0 + \mathrm{v}_d, T) / T_{R2}(\mathrm{v}_0 + \mathrm{v}_d, T), \tag{3}$$

where ν_0 and ν_d are the system frequency and Doppler shift when there is no wind, respectively, and T is the atmospheric temperature.

To obtain the vertical profile of atmospheric horizontal wind speed, it is necessary to have measurements from at least three or more directions to synthesize the radial wind speeds. This is because the atmospheric wind field is three-dimensional, and the horizontal wind speed and direction vary spatially. By measuring wind speeds from multiple directions, information about the wind speed in different directions can be obtained, allowing for the construction of the wind speed distribution in the vertical profile of the atmosphere. Assuming that the vertical wind speed can be neglected compared to the horizontal wind speed, the radial wind speed can be considered as the radial projection of the horizontal wind field. By measuring the radial wind speed in each direction, it is possible to derive the horizontal wind speed and direction through vector synthesis. There are generally two methods for vector synthesis: the three-beam, fixed-point scanning method and the four-beam, fixed-point scanning method. Using these three wind speed components, the vertical profile of wind speed in space can be constructed. Figure 3 illustrates the principle diagram of the four-beam scanning of radial wind speed and the decomposition diagram of the wind field in the radial direction. The system described in this paper utilizes the four-beam, fixed-point scanning method to measure the radial wind speed in the east, west, south, and north directions, and the zenith angle is set to 22° during actual measurement. The horizontal wind speed is decomposed into three directions: east–west (X-axis), north–south (Y-axis), and vertical (Z-axis). This approach allows for an independent consideration of the variation of wind speed in different directions, enabling a more comprehensive description and analysis of the wind field characteristics in the atmosphere.

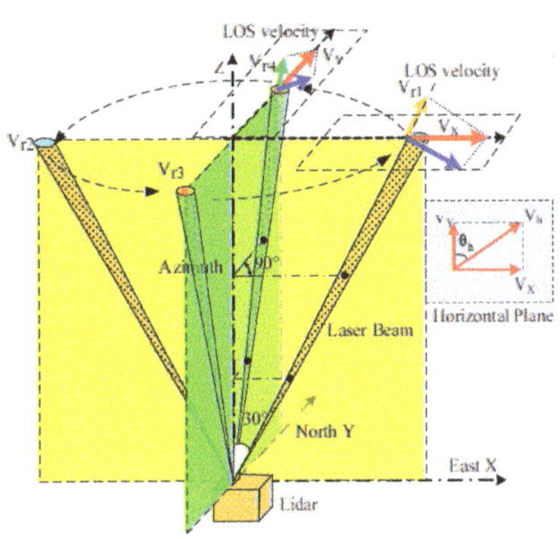

Figure 3. The principle diagram of radial wind speed was obtained using four-beam scanning.

In summary, the atmospheric horizontal wind speed and direction can be expressed as:

$$\begin{cases} V_h = \sqrt{V_x^2 + V_y^2} \\ \theta = \begin{cases} \mathrm{ar}\cos(V_y/V_h) & V_x \geq 0 \\ 2\pi - \mathrm{ar}\cos(V_y/V_h) & V_x \geq 0 \end{cases} \end{cases}, \tag{4}$$

2.4. RRDWL System Parameters

The RRDWL consists of three key components: the transmitting optics, receiving optics, and control and acquisition subsystems, as shown in Figure 4. The transmitting optics subsystem comprises a seed injection Nd:YAG solid-state laser, a laser beam expander, and the emission light path. The seed injection Nd:YAG laser generates the laser pulse, while the laser beam expander enlarges the laser beam to the desired size. The laser light is then directed through the emission light path, passing through the emission mirror and into the atmosphere.

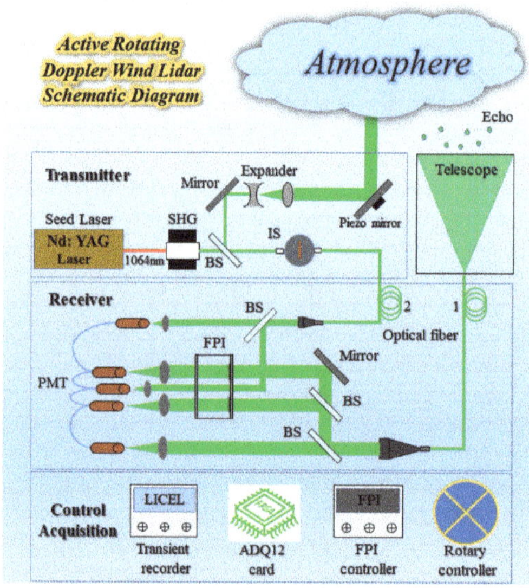

Figure 4. The principle diagram of RRDWL.

The receiving optics subsystem primarily consists of a Nessler Cassegrain receiving telescope, a receiving fiber, an FPI, a collimating device, a splitter path, interference filters, and a photomultiplier tube (PMT) among other key components. The receiving telescope captures the atmospheric backscattered light signal and focuses it onto the receiving fiber for transmission to the subsequent optical receiver. The collimating device ensures beam parallelism and stability. The echo light signal passes through the spectral light path and enters the FPI's two edge channels. Interference filters are employed to select specific wavelength ranges according to the experimental requirements. Finally, the PMT is utilized for light signal collection.

The control and acquisition subsystem includes a transient recorder, such as Licel, a high-speed acquisition card, an industrial computer, and control inversion software. These components are responsible for controlling the system's operation and parameter settings. Additionally, they facilitate the collection and processing of optical signal data.

By integrating these three subsystems, the system can effectively transmit the laser pulse into the atmosphere, receive the backscattered light signal, and control and acquire the optical data for further analysis and interpretation.

3. RRDWL Performance Test

3.1. Performance Test Result

The RRDWL was completed in October 2021, as mentioned in Refs. [25,26] provide detailed information regarding the development and parameters of the system. Therefore, this article does not go into detail. Before the RRDWL can be deployed for atmospheric wind field observation missions, it is crucial to conduct field tests to comprehensively evaluate its detection performance. These tests are conducted to ensure that the system can operate reliably in real environments and provide accurate measurements of the wind field. For this purpose, the RRDWL was deployed on Science Island, located at coordinates 117.18° E and 31.47° N, in Hefei, Anhui Province. By conducting field tests in this area, the system's capabilities can be thoroughly evaluated, and any necessary adjustments or optimizations can be made to ensure its effectiveness in practical wind field observations.

To mitigate the influence of complex weather conditions on the experimental results of the RRDWL system, the comparison observations between RRDWL and the Sonde wind field were conducted during the evening when the weather conditions were relatively favorable. It is important to acknowledge that the altitude region below 20 km in the troposphere is characterized by various weather phenomena and air turbulence. These factors can significantly amplify noise and pose challenges to the system's ability to receive stable and reliable atmospheric echo signals effectively. At the same time, the pulse repetition rate of the main laser is 30 Hz, and RRDWL needs to complete four radial wind speed detections to obtain a horizontal wind field. Therefore, in order to ensure accurate measurements, the Rotational-Rate Dual-Wavelength LiDAR (RRDWL) emits a total of 8000 laser pulses in each direction during the detection process. This large number of pulses helps compensate for the potential noise and disturbances present in the lower troposphere. By carefully considering these factors and optimizing the system's parameters, the RRDWL is able to acquire atmospheric wind field information in as little as 25 min. To enhance the signal-to-noise ratio of the system, spatial averaging processing was applied to the collected raw data, resulting in a range resolution of 300 m. In order to ensure the effective detection of atmospheric echo optical signals, the signal gate opening time delay after the completion of laser pulse Q-modulation was set to 80 microseconds. This configuration implies that the atmospheric echo optical signal profile becomes effective from a range of 11 km. Simultaneously, a weather balloon carrying a radiosonde was launched from the same location. The balloon reached an altitude of 28.01 km within 69 min, covering the entire operating time of the LiDAR system. According to GPS data, the balloon drifted 165.6 km to the east during its ascent.

In Figure 5a, the horizontal wind speed (WS) profiles obtained using the RRDWL and Sonde are depicted using black dots and red lines, respectively. In Figure 5b, the absolute deviation between the RRDWL and Sonde measurements is shown. It is observed that, except for a thin layer around 13 km and 18 km, the absolute deviation of the wind speed at other altitudes remains below 10 m/s. At a height of 12 km near the tropopause, the horizontal wind speed reaches a maximum of 90 m/s and then gradually decreases to within 5 m/s at an altitude of 22 km. Between 22 km and 27 km, a weak layer of wind is observed with peak winds of approximately 17 m/s. Overall, the comparison between the RRDWL and Sonde measurements shows reasonably good agreement, with the absolute deviation of wind speed mostly below 10 m/s, except for certain altitudes. In Figure 5c, the wind direction (WD) profile obtained using both the RRDWL and Sonde is presented. It can be observed that westerly winds, ranging from 227° to 292°, are prevalent at altitudes between 10 km and 27 km. This indicates a predominant westward flow in the atmosphere within this altitude range. At approximately 28 km, both the RRDWL and radiosonde measurements indicate the presence of wind shear. In Figure 5d, the deviation of wind direction between the RRDWL and Sonde measurements is shown. It is noteworthy that the deviation of wind direction between the RRDWL and Sonde measurements remains below 10° below 25 km. This suggests a relatively good agreement between the two devices in capturing the wind direction profile within this altitude range.

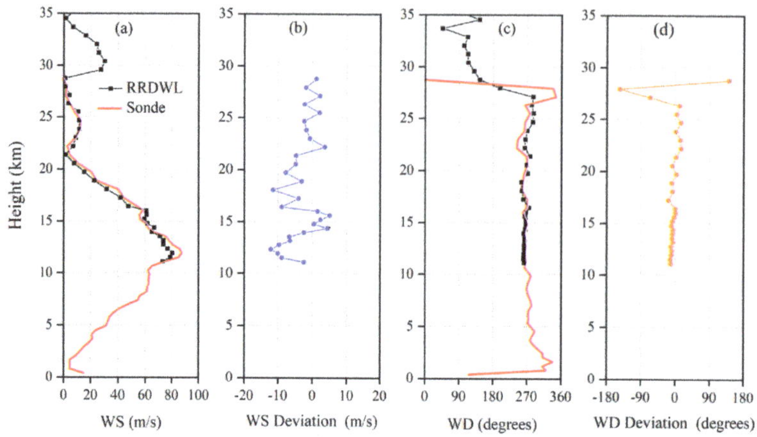

Figure 5. Comparison of horizontal wind profiles obtained using RRDWL and Sonde. (**a**) wind speed (**b**) deviation of WS (**c**) wind direction (**d**) deviation of WD.

3.2. Measurement Error Analysis

During actual atmospheric wind field detection experiments, it is common for the detected data to deviate from theoretical simulation results due to measurement errors. These errors can be categorized into systematic errors and random errors, each contributing differently to the overall measurement accuracy and precision. Systematic errors are typically used to describe the accuracy of measurement results, representing consistent biases or offsets in the measurements. On the other hand, random errors are used to describe the precision of the measurements, reflecting variations or fluctuations in the measured values. The measurement accuracy, which quantitatively expresses the influence of both types of errors on the measurement results, can be characterized by uncertainty. The impact of these two types of errors on the measurement results is illustrated by the position distribution of red dots in Figure 6. Taking the small figure within Figure 5 as an example, Figure 6a indicates high measurement accuracy and small uncertainty, suggesting low systematic and random errors. In contrast, Figure 6d represents low measurement accuracy and large uncertainty, indicating significant systematic and random errors. Figure 6b demonstrates common errors encountered during detection, such as those caused by improper calibration coefficients. However, Figure 6c depicts the typical measurement errors of LiDAR, primarily dominated by random errors stemming from optical quantum noise and errors associated with system parameters. When both systematic and random errors are minimized, measurement results similar to those shown in Figure 6a can be obtained. The systematic errors of the RRDWL primarily arise from laser linewidth jitter, laser frequency jitter, system calibration, and inversion methods. These factors contribute to the overall systematic error in the measurements obtained using the system.

The signal-to-noise ratio (SNR) is a crucial metric widely used to assess the effective detection height in LiDAR systems. It quantifies the ratio of the desired signal strength to the level of background noise present in the received signal. In general, a higher signal intensity results in a higher SNR. SNR plays a significant role in determining the appropriate inversion height and calibration point in LiDAR systems. It directly influences the accuracy and reliability of the inversion results for atmospheric parameters. By selecting an inversion height with a higher SNR, the system can obtain more reliable and accurate measurements of atmospheric parameters. The formula to calculate SNR is typically defined as the ratio of the signal power to the noise power. However, the specific formulation can vary depending on the system and measurement configuration. It is essential to consider the characteristics and requirements of the LiDAR system in determining the appropriate SNR calculation method. Specifically, the SNR of RRDWL can be calculated as follows:

$$SNR(z) = \frac{N_X(z)}{\sqrt{N_X(z) + 2(N_b + N_d)}} \cdot \sqrt{n}, \quad (5)$$

where N_X (Z) is the atmospheric echo signal, N_b and N_d are the backscattered noise and atmospheric background radiation noise, and n is the pulse accumulation number.

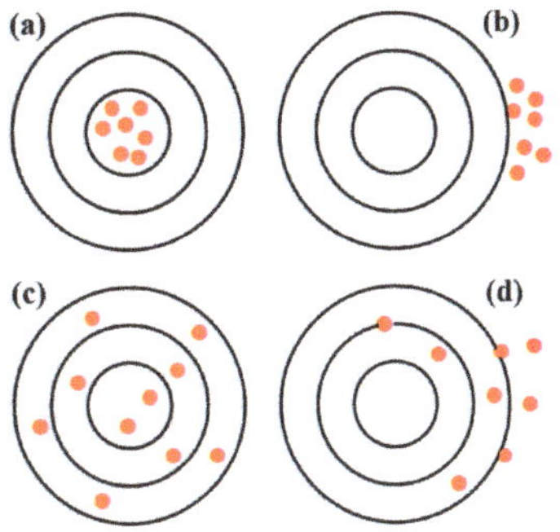

Figure 6. Schematic diagram of random and systematic errors in measurement. (**a**) Small systematic error and random error (**b**) Large systematic error and small random error (**c**) Small systematic error and large random error (**d**) Both systematic error and random error are large.

In general, a region is considered valid if its signal-to-noise ratio (SNR) exceeds a predetermined threshold, such as 10. Consequently, a set of test data was selected to analyze the SNR profiles of the three channels of a four radial FPI. Figure 7 illustrates the results, indicating that the measured SNR is approximately 112 at a height of 30 km and around 10 at a height of 42 km. However, it is observed that the four radial signals gradually become unstable, starting from an altitude of 30 km, with the SNR exhibiting increasing jitter. It is important to note that, during actual measurements, several factors can cause inconsistencies in the SNR of different radials at the same height. These factors include the uneven distribution of the atmosphere and differences in the efficiency of the system's detector components. Such variations can affect the strength of the signal and the level of noise, consequently resulting in disparities in the SNR values among different radials.

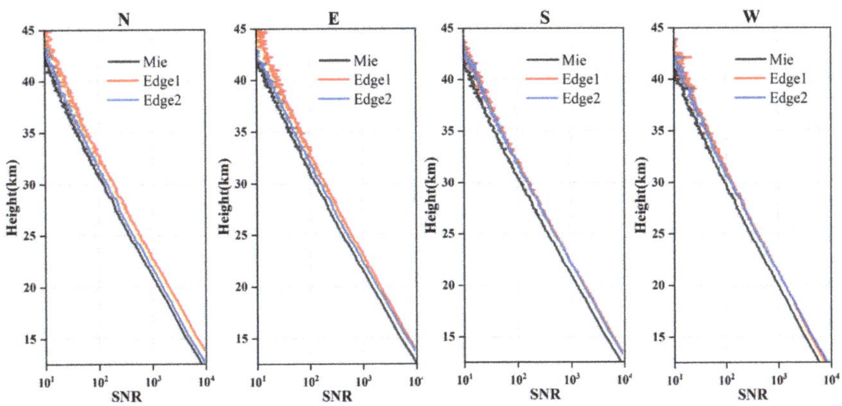

Figure 7. Four radial SNR profiles of a set of test data.

To evaluate the effective altitude of the system in long-term detection, an experiment was conducted wherein 8000 pulses were transmitted, and an SNR threshold of 10 was used as the evaluation criteria. The signals obtained during favorable atmospheric conditions were statistically analyzed, and the results are presented in Figure 8. In Figure 8a, the statistics for the effective detection heights of the two edge channels are displayed. It can be observed that edge channel 1 has a total of 17 sets of effective heights between 38 km and 40 km, and 26 sets between 40 km and 45 km. On the other hand, edge channel 2 has 12 sets with effective heights between 38 km and 40 km and 31 sets with effective heights between 40 km and 45 km. Figure 8b shows the statistical results of the combined effective detection altitudes. There are a total of 16 groups between 38 km and 40 km and 27 groups between 40 km and 43 km. In summary, the analysis indicates that the effective detection height of the RRDWL is primarily concentrated between 38 km and 42 km.

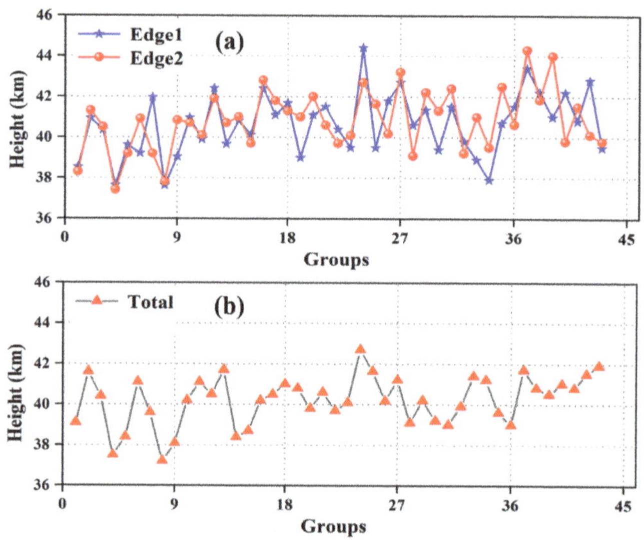

Figure 8. Effective detection height sequence of RRDWL. (**a**) edge channel (**b**) system overall.

When using atmospheric backscatter signals with errors to invert atmospheric wind speed, the accuracy of the results can be affected by the transmission effect of the errors. The main source of measurement errors consists of random factors such as optical quantum noise, and it is closely associated with the technical parameters of the LiDAR system. During actual detection experiments, the presence of system errors can lead to the offset of the transmitting laser frequency from the intersection of the transmittance curve of the edge channel of the FPI. This offset can result in lower frequency locking accuracy. The wind speed error at this time is derived, and the frequency response function is:

$$R_V = \frac{N_1}{N_2}, \tag{6}$$

where N_1 and N_2 indicate the edge channel transmittance of FPI. Through the error transfer function, the error in calculating radial wind speed is:

$$\varepsilon = \frac{1}{\theta(S/N)}, \tag{7}$$

According to the mean square error of the actual frequency response function:

$$\begin{aligned}\frac{\delta R_V}{R_V} &= \sqrt{(\frac{\partial \ln R_V}{\partial N_1}\delta N_1)^2 + (\frac{\partial \ln R_V}{\partial N_2}\delta N_2)^2} \\ &= \sqrt{(\frac{\delta N_1}{N_1})^2 + (\frac{\delta N_2}{N_2})^2} \\ &= \sqrt{(\frac{1}{SNR_1})^2 + (\frac{1}{SNR_2})^2}\end{aligned}, \tag{8}$$

Therefore, the actual wind speed error can be obtained according to the wind speed sensitivity and error transfer function:

$$\theta_V = \frac{\partial R_V/\partial V}{R_V}, \tag{9}$$

$$(\frac{\delta R_V}{R_V})^2 = (\frac{\partial \ln R_V}{\partial V}\delta V)^2 = (\theta_V \delta V)^2, \tag{10}$$

$$\partial V = \frac{\frac{\delta R_V}{R_V}}{\theta_V} = \frac{\sqrt{(\frac{1}{SNR_1})^2 + (\frac{1}{SNR_2})^2}}{\theta_V} \tag{11}$$

To provide a more representative error analysis, another set of experimental data was utilized, and the random errors of the detected wind profiles were calculated using the aforementioned formula. The results are presented in Figure 9. In Figure 9a, the random errors of the four radial wind profiles in the east, south, west, and north directions are depicted. Figure 8b displays the random errors of the combined X and Y directions, as well as the horizontal wind speeds. According to the results shown in the figure, the random error of the wind profile exhibits an exponential increase with altitude. At 36 km, the four radial wind profiles demonstrate similar random error characteristics, with an approximate value of 10 m/s. The random error of the resulting horizontal wind speed is around 10 m/s at 34 km. The exponential growth of random errors with height emphasizes the importance of exercising caution when analyzing wind profile data at higher altitudes. Additionally, the similarity in random error characteristics among the four radial wind velocity profiles indicates consistency in data acquisition and processing. On the other hand, the random error of the synthesized horizontal wind velocity profile remains at a relatively high level. This finding underscores the need to fully consider the influence of random errors on the synthesis and inversion of horizontal wind velocities.

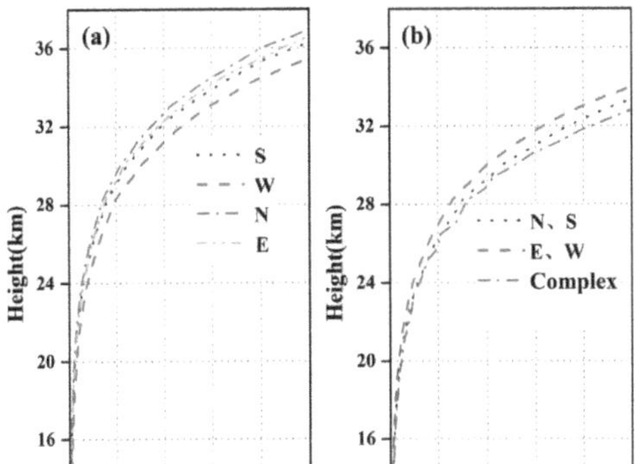

Figure 9. Random error of wind speed at different altitudes (**a**) radial (**b**) synthesis.

To provide a more intuitive comparison of the horizontal wind speed error profiles generated using the RRDWL, Figure 10 displays a diagram showcasing these profiles.

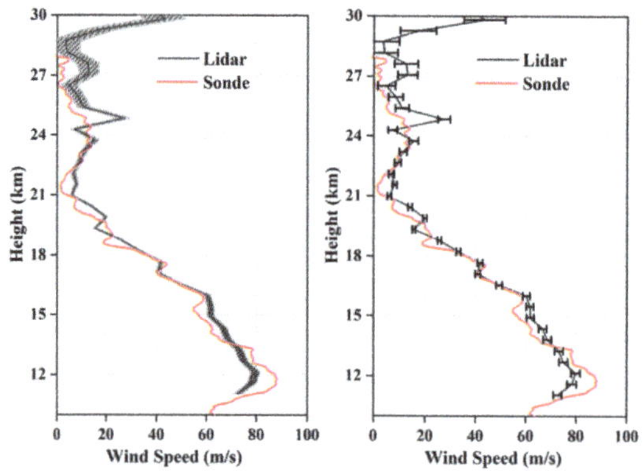

Figure 10. A set of horizontal wind speed error profiles for RRDWL.

To compare the horizontal wind profiles obtained using the RRDWL and the Sonde, as well as the random error of wind speed calculated using SNR, Figure 11 provides a visual representation. In Figure 11a, the distribution trend of the horizontal wind speed profiles obtained using both devices is shown to be essentially the same. In Figure 11a, the distribution trend of the horizontal wind speed profiles obtained using both devices is shown to be essentially the same. The variation trend of the measured wind speed error and random error is similar to the height changes. This finding validates that the random error calculated using SNR can, to some extent, represent the wind measurement error and can be used to verify the accuracy of wind measurements. Furthermore, within the height range from 15 km to 25 km, significant jitter and large amplitude of the measured error can be observed. This suggests that system errors have a considerable impact on the accuracy of measurement results, in addition to the influence of atmospheric level heterogeneity. Figure 10 also highlights the significant effect of the zenith angle of the receiving telescope on the horizontal wind profile. Initially set at 22°, it was later found that setting the zenith angle to 24° resulted in better agreement between the RRDWL and radiosonde data. This indicates that the zenith angle plays an important role in the accuracy of measurement results. However, it is important to note that the zenith angle has less impact on the random error. Random errors are typically caused by various unpredictable factors, and the zenith angle mainly affects the measurement results based on system settings and instrument characteristics. The analysis results highlight the significant impact of system errors on the accuracy of measurement results, independent of atmospheric level heterogeneity. Additionally, the zenith angle of the receiving telescope is shown to have a notable effect on the accuracy of the horizontal wind profile. These findings emphasize that while the Rotary Rayleigh Doppler Wind Lidar (RRDWL) enables single-unit detection of the horizontal wind field, the presence of systematic errors can introduce significant inaccuracies in wind speed measurements.

To enhance the representativeness of the experimental results, statistical analysis was conducted on 35 sets of data obtained from successive detection experiments. Since the detection height of the Sonde does not typically exceed 30 km, the random error sequence of the wind speed in the horizontal wind profile at a height of 30 km, obtained from the RRDWL, was calculated and plotted in Figure 12. By examining Figure 11, it can be observed that the random error of the horizontal wind speed is below 10 m/s in 11 out of the 35 groups, while it exceeds 15 m/s in 5 out of the 35 groups. On average, the random error of the horizontal wind speed amounts to approximately 11.3 m/s. In summary, the effective detection height of the RRDWL system is primarily concentrated between 38 km and 42 km. Additionally, at an altitude of 30 km, the random error of the horizontal wind speed is approximately 11.3 m/s. Taking into account the statistical analysis of multiple experiments, these findings provide valuable insights into the performance and limitations of the RRDWL system. They contribute to understanding the effective detection range of the system and the expected level of random error in horizontal wind speed measurements at different altitudes.

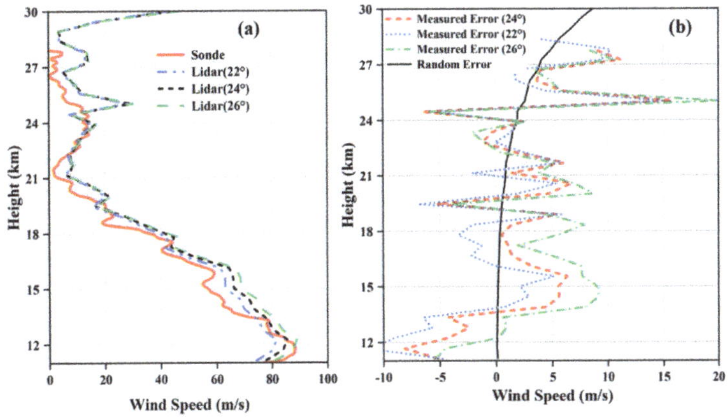

Figure 11. Horizontal wind field distribution under different zenith angles: (**a**) Wind profile comparison. (**b**) Measured error and random error.

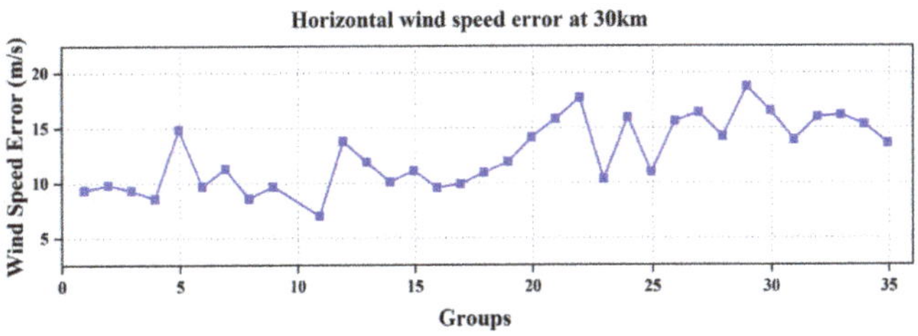

Figure 12. Continuous detection of 30 km horizontal wind speed error sequence.

4. Error Tracing Analysis

According to Section 3.2, the detection results of the RRDWL system are primarily influenced by two types of errors. The first type is a random error, primarily originating from light quantum noise. The second type of error arises from system-related factors, such as laser frequency jitter, operational stability, calibration, and inversion methods. Therefore, conducting traceability analysis to identify the main source of system error is crucial for optimizing the system's performance. Considering that the RRDWL system utilizes a single unit for integral rotation, weighing nearly 6 tons, it is inevitable to encounter stability issues in the transceiver system during the actual detection process due to vibrations and structural deformations. These factors can significantly impact the overall detection performance. As depicted in Figure 13, on the evening of 13 March 2022, a continuous detection test was conducted by the system. Before the wind field detection experiment that night, a pre-experiment was carried out to observe the signal-to-noise ratio due to the cloudy weather, and no gating signal was set. The comparison diagram of atmospheric echo signals from three consecutive cycles in the north, east, south, and west (following the rotation sequence) reveals significant deviations within each group of four radial atmospheric echo signals over a short period of time (approximately 90 min), with varying change patterns. Based on this observation, it can be inferred that these deviations are caused by uneven ground conditions and that the deformation of the shelter is due to rotation, subsequently disrupting the collimation of the optical path in the transceiver subsystem.

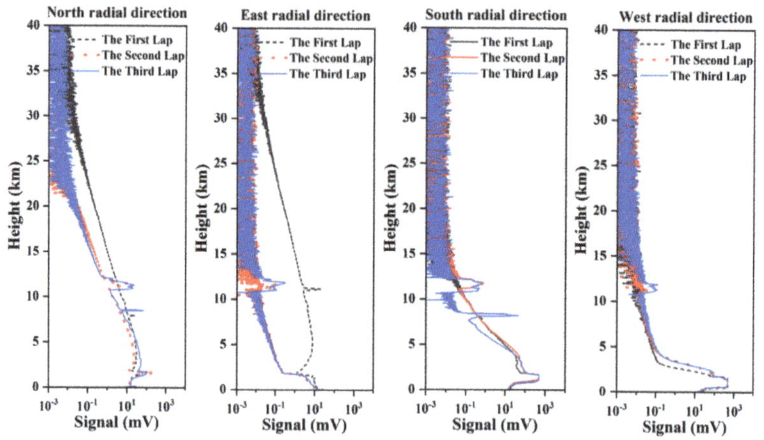

Figure 13. The change of four-radial atmospheric echo signal during continuous rotation.

In the emission subsystem, ensuring the stability of optical devices such as the fiber seed laser, main laser, and laser beam expander is crucial to maintaining the quality and frequency of the emitted laser beam. As stated, the equipment is in an open state during shelter operation, and it is inferred that the deformation of the resonator is primarily caused by changes in ambient temperature. To investigate this issue further, an ambient temperature test experiment was conducted in the afternoon of 19 January 2022, during winter, as shown in Figure 14. The purpose of this experiment was to assess the impact of ambient temperature on the resonator's deformation. Initially, the main laser exhibited a fundamental frequency energy of approximately 1.5 J with a good pulse shape, indicating single longitudinal mode oscillation. After being turned off for four days, the factors causing changes in the main laser's performance were investigated experimentally on January 24. Between 10:00 a.m. and 11:30 a.m., the equipment shelter was closed, and the main laser was preheated for 30 min. Despite the air conditioning being in heating mode, the 532 nm laser pulse exhibited a poor shape, showing a multi-longitudinal pattern, and the pulse energy was significantly reduced to around 300 mJ. This phenomenon was attributed to the change in ambient temperature. The room temperature at this time was 27 °C. Interestingly, when the air conditioner was turned off, the pulse shape improved, and the pulse energy was restored to about 620 mJ by adjusting the rear mirror. At 17:30 p.m., the main door of the equipment shelter was opened, causing the room temperature to rapidly drop from 27 °C to 20 °C. However, there were no significant changes observed in the pulse energy and waveform at this point, indicating that the cooling effect was not sufficient. To prolong the cooling time, the decision was made to close the hatch door. By 20:40 in the evening, the room temperature had reached 19 °C. Upon restarting the main laser, it was discovered that the energy of the fundamental amplification stage after seed injection had decreased to 0.74 J, while the energy of the 532 nm pulse had sharply dropped to 178 mJ. Subsequently, both the main and side doors of the equipment shelter were opened, causing the room temperature to rapidly decrease from 19 °C to 13 °C, allowing for the observation of the secondary effects of ambient temperature changes on the main laser. At this point, the energy of the 532 nm pulse was reduced to 160 mJ, and the pulse shape became highly chaotic, exhibiting a bimodal structure. At 21:10, the cabin door was closed, and the air conditioning was switched back to heating mode, causing the temperature to rise rapidly from 12 °C to 23 °C. However, the pulse energy of the oscillatory stage did not change significantly, and the waveform remained disordered. Efforts were made to strengthen the heating effect of the air conditioner, and, after working for more than half an hour, the room temperature was restored to 26 °C. However, the fundamental frequency oscillation stage energy after seed injection was only 107 mJ, indicating that the drastic change in ambient temperature had caused irreversible damage to the main laser and the seed injection process. As a result of this irreversible damage, the quality of the laser beam emitted could not be restored to normal levels.

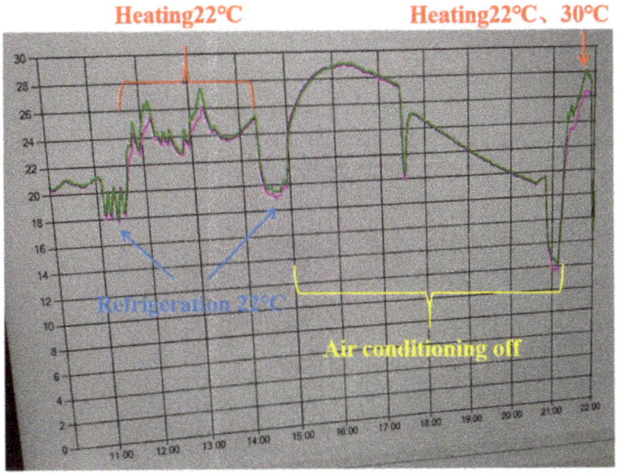

Figure 14. Main laser ambient temperature change curve.

This experimental investigation underscores the sensitivity of the main laser to ambient temperature changes and highlights the irreversible damage that can occur when subjected to rapid and extreme temperature fluctuations. Implementing measures to stabilize the ambient temperature and ensure gradual temperature transitions can help mitigate such issues and preserve the optimal performance of the main laser in the RRDWL system.

In the receiving subsystem, the stability of the optical receiver, particularly the successor optical receiver, is crucial. The transmittance curve of the FPI channel, which is incident upon by the beam, can be described using an Airy line, regardless of the beam's divergence angle. This property offers significant advantages when measuring transmittance curves, as only a small number of points need to be measured to accurately fit the entire curve. By reducing the number of measurement points, the system can accumulate more laser pulses for each data point, thereby improving the signal-to-noise ratio. This enables more accurate measurements without requiring excessive scanning time. However, during the actual process of scanning the transmittance, it has been observed that when the beam has a divergence angle (which is usually unavoidable) and oblique incidence on the FPI, the transmittance curve becomes asymmetric. In such cases, accurately describing the transmittance curve using analytical formulas becomes challenging. To obtain an accurate spectral line of the FPI transmittance, a very small scanning step size is required to reduce the number of measured pulses at each point. Even with this measure, it takes nearly 40 min to accurately scan the entire FPI curve. Therefore, a good solution before scanning the transmittance curve of the FPI is to adjust the beam so that it is directly on the FPI without any divergence angle. However, in the early design of the receiver, the influence of the position and angle of the optical elements (such as filters, beam-splitting prisms, etc.) in the front and back optical paths of the FPI on mirror reflection was not fully considered. Additionally, due to vibrations caused by continuous rotation, the relative position of the optical elements inside the receiver may slightly change. The incomplete parallelism between the optical element and the surface of the FPI increases the degree of non-normal incidence of the beam. As a result, the detected signal contains components of oblique incidence through the FPI and surface reflection after the FPI, making it almost impossible to accurately describe the transmittance spectral line of the FPI using an analytical expression. Figure 15 displays the FPI transmittance spectral lines obtained from several rotation-detection experiments, along with the FPI transmittance spectral lines under ideal conditions. One can observe that, despite adjusting the incident beam to normal incidence, there remains a slight asymmetry in the spectral lines. However, after conducting numerous rotational detection tests, the spectral lines exhibit a significant increase in asymmetry. Upon inspecting the receiver, it was discovered that the root of the problem was the shift in the relative position of the mirror after the FPI, causing the incident light to reflect when passing through the filter in front of the photomultiplier tube (PMT). To address this issue, careful consideration should be given to the design of the receiver, taking into account the position and angle of the optical elements in relation to the FPI. Additionally, measures should be implemented to minimize vibrations and ensure the stability of the optical elements within the receiver. By addressing these factors, the accuracy of the transmittance measurements and the spectral line description of the FPI can be improved, leading to enhanced performance and reliability of the receiving subsystem.

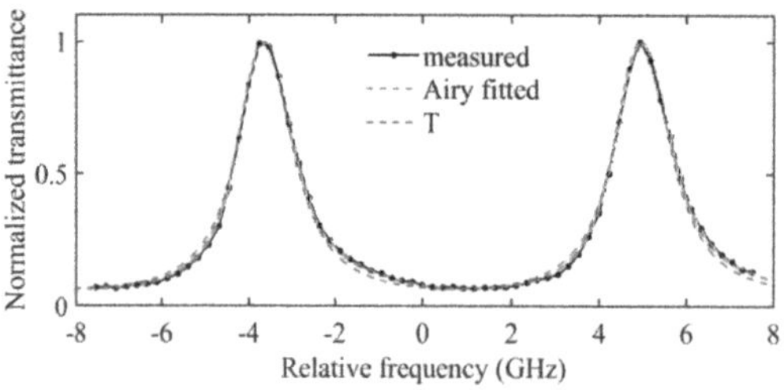

Figure 15. Influence of non-normal incidence and specular reflection of incident light on spectral lines of transmittance of FPI.

5. Discussion

The atmospheric wind field data from the upper troposphere to the stratosphere (10–50 km) is of great significance for the in-depth study of the dynamics of the middle and upper atmosphere and the interaction between different layers. Compared with other detection methods, direct wind LiDAR has significant advantages in spatial resolution, detection range, and flexibility. At the same time, it is the only detection technology that can continuously and effectively obtain the high-precision data of three-dimensional wind fields in the troposphere and even the upper atmosphere.

Traditional direct wind LiDAR systems can be classified into two types. The first type is the scanning type, which incorporates a scanning mirror composed of two flat mirrors positioned above the LiDAR's transmitter and receiver. This configuration enables the LiDAR system to detect the horizontal wind field by utilizing cone scanning. However, when the aperture of the receiving telescope exceeds half a meter, the size, thickness, and weight of the scanning mirror increase significantly. This not only poses challenges in terms of manufacturing complexity and cost but also makes optical tuning more difficult. Consequently, this method is only suitable for small and medium-sized systems, limiting the effective detection height of the system to the upper troposphere. Alternatively, researchers have employed a different approach to detect high-altitude wind fields by utilizing two (or three) large-aperture LiDAR systems arranged orthogonally (or at fixed angles) to measure different radial wind speeds. Typically, each system in this configuration has the same configuration and system parameters. By simultaneously measuring the line-of-sight wind speeds in different directions, it becomes possible to obtain the atmospheric level wind field at stratospheric and potentially higher altitudes. This detection method offers significant advantages, such as higher spatiotemporal resolution and an increased detection altitude range. However, it also presents certain challenges. Firstly, employing two or more subsystems necessitates the use of at least two high-power pulsed lasers, multiple large-aperture telescopes, and multiple detection and acquisition systems. This leads to increased development and maintenance costs while also raising the difficulty of precisely calibrating the instruments. Additionally, appropriate algorithms need to be employed to mitigate the impact of measurement discrepancies arising from errors in different subsystems. Since each subsystem may introduce varying sources of error, such as deviations in optical calibration or differences in detector response, these errors can potentially affect the final measurement results of the wind field. More importantly, whether it is a scanning system or a large aperture orthogonal system, the detection performance of direct wind LiDAR largely depends on the accuracy of the instrument. This includes factors such as the quality of the laser beam, the accuracy of the discriminator, and the response of the detector. The accuracy of these key parameters is crucial to obtain accurate and reliable wind field data. In addition, due to the need for larger telescopes and additional system configurations, the development cost of direct wind lidar is higher than that of coherent wind lidar. This means the direct wind LiDAR may be limited in practical applications. Another challenge is that different direct wind LiDAR systems use different discriminators and scanning modes, leading to differences in calibration methods. Calibration is an important step to ensure the measurement accuracy of the system. Different calibration methods may lead to inconsistencies between subsystems and limit the comparison and fusion of data.

In summary, the key factors that impact the performance of direct wind LiDAR can be further categorized as follows:

1. Laser Frequency Drift: Direct wind LiDAR systems typically use seed injection lasers, which can experience frequency drift due to temperature variations or vibrations. This drift can lead to significant errors in wind speed inversion.
2. Laser Beam Divergence and Stability: In real-world environments, the emitted laser beam may exhibit jitter or amplification in terms of divergence angle and frequency stability. This can impede the effective identification of Doppler frequency shifts caused by wind, resulting in increased errors in wind speed inversion. Suitable algorithms and devices are necessary to accurately identify and lock onto the frequency.
3. Instrument Stability and Calibration: Direct wind LiDAR systems often consist of multiple sets of precision instruments. The complex working environment can introduce varying degrees of error, impacting the stability and calibration of these instruments.

These factors collectively contribute to the slower research progress in direct wind LiDAR for detecting atmospheric wind fields. Additionally, it is worth noting that most field detections of direct wind LiDAR systems, both domestically and abroad, are primarily concentrated in high-latitude areas or regions with favorable meteorological conditions. This indicates that different meteorological conditions and complex external field environments present challenges for direct wind LiDAR, requiring further exploration and research.

Based on the above background, this research group has developed the first rotary Rayleigh Doppler wind LiDAR system in China. This innovative system utilizes a rotating platform structure to enable the detection of four radial wind speeds through integral rotation. The core components of the system include a seed injection laser, a large aperture receiving telescope, and a successor optical receiver. This design not only overcomes the limitation of traditional large aperture systems, which can only measure a single radial wind speed, but also enables a single direct wind measurement LiDAR system to detect the horizontal wind field in the middle and upper atmosphere. As a result, the system achieves a higher level of wind field detection compared to conventional approaches. Additionally, this design reduces the complexity of the structure and effectively lowers the development cost of the equipment. This development represents a significant advancement in direct wind LiDAR technology, offering improved capabilities for measuring wind speeds in different directions simultaneously. By utilizing the rotary Rayleigh Doppler wind detection LiDAR system, researchers can obtain comprehensive wind field data in the middle and upper atmosphere. This achievement holds promise for expanding the applications of direct wind LiDAR systems and advancing atmospheric research.

6. Conclusions

In order to effectively observe the stratospheric atmospheric wind field in the high troposphere, Xie et al. from the Chinese Academy of Sciences, Anhui Institute of Optics and Fine Mechanics (CAS, AIOFM) developed an innovative rotary Rayleigh Doppler wind measurement LiDAR (RRDWL). Following the completion of system construction, a performance test experiment was conducted to verify the feasibility of monomer detection of the atmospheric wind field in real-world environments. The results demonstrated that the system was capable of detecting the atmospheric wind field within an altitude range from 38 km to 42 km, with a horizontal wind speed error of approximately 11.3 m/s at 30 km altitude.

Furthermore, error analysis was performed to investigate the main factors influencing the system's performance. It was identified that RRDWL had limitations in long-term continuous detection. The primary factor contributing to the reduction in effective detection height was the damage to optical axis parallelism caused by vibrations introduced by the rotating shelter and uneven ground surfaces. Additionally, random fluctuations in ambient temperature disrupted the seed laser injection process, thereby affecting the quality of the emitted beam. The imbalance in FPI parallelism resulting from non-normal incident beams and specular reflections had a significant impact on the accuracy of frequency discrimination. These error-tracing analyses provided valuable insights for optimizing the performance of RRDWL. Addressing the issues related to optical axis parallelism, vibration control, ambient temperature stability, seed laser injection, and FPI parallelism imbalance will be crucial for enhancing the system's accuracy and reliability. Moreover, these findings serve as a foundation for the future long-term continuous supply of mid-upper atmospheric wind field data using RRDWL.

Author Contributions: Supervision, conceptualization, C.X.; resources, K.X.; software, B.W.; methodology, J.J. and L.L.; numerical simulation, writing—original draft preparation, J.C.; writing—review and editing, M.Z. All authors have read and agreed to the published version of the manuscript.

Funding: This work was supported in part by the Strategic Priority Research Program of the Chinese Academy of Sciences (XDA17040524), Anhui Province science and technology major project (201903c08020013).

Data Availability Statement: The data mentioned in the manuscript may be requested by email from the author.

Acknowledgments: I would like to thank the National Meteorological Center of China for the radiosonde data.

Conflicts of Interest: The authors declare no conflicts of interest.

References

1. Yang, J.; Xiao, C.; Hu, X.; Xu, Q. Responses of zonal wind at ~40° N to stratospheric sudden warming events in the stratosphere, mesosphere and lower thermosphere. *Sci. China Technol. Sci.* **2017**, *60*, 935–945. [CrossRef]
2. Zhao, R.; Dou, X.; Xue, X.; Sun, D.; Han, Y.; Chen, C.; Zheng, J.; Li, Z.; Zhou, A.; Han, Y.; et al. Stratosphere and lower mesosphere wind observation and gravity wave activities of the wind field in China using a mobile Rayleigh Doppler lidar. *J. Geophys. Res. Space Phys.* **2017**, *122*, 8847–8857. [CrossRef]
3. Wang, Z.; Liu, Z.; Liu, L.; Wu, S.; Liu, B.; Li, Z.; Chu, X. Iodine-filter-based mobile Doppler lidar to make continuous and full-azimuth-scanned wind measurements: Data acquisition and analysis system, data retrieval methods, and error analysis. *Appl. Opt.* **2010**, *49*, 6960–6978. [CrossRef] [PubMed]
4. Hedin, A.E.; Fleming, E.L.; Manson, A.H.; Schmidlin, F.J.; Avery, S.K.; Clark, R.R.; Franke, S.J.; Fraser, G.J.; Tsuda, T.; Vial, F.; et al. Empirical wind model for the upper, middle and lower atmosphere. *J. Atmos. Terr. Phys.* **1996**, *58*, 1421–1447. [CrossRef]
5. Prasad, N.S.; Sibell, R.; Vetorino, S.; Higgins, R.; Tracy, A. All-Fiber, Modular, Compact Wind Lidar for Wind Sensing and Wake Vortex Applications. *Proc. SPIE* **2015**, *9465*, 94650C.
6. Dou, X.; Han, Y.; Sun, D.; Xia, H.; Shu, Z.; Zhao, R.; Shangguan, M.; Guo, J. Mobile Rayleigh Doppler lidar for wind and temperature measurements in the stratosphere and lower mesosphere. *Opt. Express* **2014**, *22*, A1203–A1221. [CrossRef] [PubMed]
7. Han, F.; Liu, H.; Sun, D.; Han, Y.; Zhou, A.; Zhang, N.; Chu, J.; Zheng, J.; Jiang, S.; Wang, Y. An Ultra-narrow Bandwidth Filter for Daytime Wind Measurement of Direct Detection Rayleigh Lidar. *Curr. Opt. Photonics* **2020**, *4*, 69–80.
8. Lombard, L.; Valla, M.; Planchat, C.; Goular, D.; Augère, B.; Bourdon, P.; Canat, G. Eyesafe coherent detection wind lidar based on a beam-combined pulsed laser source. *Opt. Lett.* **2015**, *40*, 1030–1033. [CrossRef] [PubMed]
9. Liu, H.; Zhu, X.; Fan, C.; Bi, D.; Liu, J.; Zhang, X.; Zhu, X.; Chen, W. Field Performance of All-Fiber Pulsed Coherent Doppler Lidar. *Eur. Phys. J. Conf.* **2020**, *237*, 08009. [CrossRef]

10. Gentry, B.M.; Chen, H.; Li, S.X. Wind measurements with 355-nm molecular Doppler lidar. *Opt. Lett.* **2000**, *25*, 1231–1233. [CrossRef]
11. Feifei, L.; Decang, B.; Heng, L.; Lucheng, Y.; Mingjian, W.; Xiaopeng, Z.; Jiqiao, L.; Weibiao, C. Principle Prototype and Experimental Progress of Wind Lidar in Near Space. *Chin. J. Lasers* **2020**, *47*, 0810003. [CrossRef]
12. Souprayen, C.; Garnier, A.; Hertzog, A. Rayleigh-Mie Doppler wind lidar for atmospheric measurements. II. Mie scattering effect, theory, and calibration. *Appl. Opt.* **1999**, *38*, 2422–2431. [CrossRef] [PubMed]
13. Chanin, M.L.; Garnier, A.; Hauchecorne, A.; Porteneuve, J. A doppler lidar for measuring winds in the middle atmosphere. *Geophys. Res. Lett.* **1989**, *16*, 1273–1276. [CrossRef]
14. Baumgarten, G. Doppler Rayleigh/Mie/Raman lidar for wind and temperature measurements in the middle atmosphere up to 80 km. *Atmos. Meas. Tech.* **2010**, *3*, 1509–1518. [CrossRef]
15. Huffaker, R.M.; Reveley, P.A. Solid-state coherent laser radar wind field measurement systems. *Pure Appl. Opt. J. Eur. Opt. Soc. Part A* **1998**, *7*, 863–873. [CrossRef]
16. Zhao, Y.; Zhang, X.; Zhang, Y.; Ding, J.; Wang, K.; Gao, Y.; Su, R.; Fang, J. Data Processing and Analysis of Eight-Beam Wind Profile Coherent Wind Measurement Lidar. *Remote Sens.* **2021**, *13*, 3549. [CrossRef]
17. Zhang, H.; Liu, X.; Wang, Q.; Zhang, J.; He, Z.; Zhang, X.; Li, R.; Zhang, K.; Tang, J.; Wu, S. Low-Level Wind Shear Identification along the Glide Path at BCIA by the Pulsed Coherent Doppler Lidar. *Atmosphere* **2020**, *12*, 50. [CrossRef]
18. Guo, W.; Shen, F.; Shi, W.; Liu, M.; Wang, Y.; Zhu, C.; Shen, L.; Wang, B.; Zhuang, P. Data inversion method for dual-frequency Doppler lidar based on Fabry-Perot etalon quad-edge technique. *Optik* **2018**, *159*, 31–38. [CrossRef]
19. Zhang, F. Research on Doppler Wind Lidar System with Wind Detection of High Temporal and Spatial Resolution. Ph.D. Thesis, University of Science and Technology of China, Hefei, China, 2015. (In Chinese).
20. Shen, F.; Xie, C.; Qiu, C.; Wang, B. Fabry-Perot etalon-based ultraviolet trifrequency high-spectral-resolution lidar for wind, temperature, and aerosol measurements from 0.2 to 35 km altitude. *Appl. Opt.* **2018**, *57*, 9328–9340. [CrossRef]
21. Xia, H.; Dou, X.; Shangguan, M.; Zhao, R.; Sun, D.; Wang, C.; Qiu, J.; Shu, Z.; Xue, X.; Han, Y.; et al. Stratospheric temperature measurement with scanningFabry-Perot interferometer for wind retrieval from mobile Rayleigh Doppler lidar. *Opt. Express* **2014**, *22*, 21775–21789. [CrossRef]
22. Liu, Z.-S.; Wu, D.; Liu, J.-T.; Zhang, K.-L.; Chen, W.-B.; Song, X.-Q.; Hair, J.W.; She, C.-Y. Low-altitude atmospheric wind measurement from thecombined Mie and Rayleigh backscattering by Doppler lidar with an iodine filter. *Appl. Opt.* **2002**, *41*, 7079–7086. [CrossRef] [PubMed]
23. Ansmann, A.; Ingmann, P.; Le Rille, O.; Lajas, D.; Wandinger, U. Particle backscatter and extinction profiling with the spaceborne HSR Doppler wind lidar ALADIN. *Appl. Opt.* **2007**, *46*, 6606–6622. [CrossRef] [PubMed]
24. Zhang, C.; Sun, X.; Zhang, R.; Zhao, S.; Lu, W.; Liu, Y.; Fan, Z. Impact of solar background radiation on the accuracy of wind observations of spaceborne Doppler wind lidars based on their orbits and optical parameters. *Opt. Express* **2019**, *27*, A936–A952. [CrossRef] [PubMed]
25. Zhao, M.; Xie, C.; Wang, B.; Xing, K.; Chen, J.; Fang, Z.; Li, L.; Cheng, L. A Rotary Platform Mounted Doppler Lidar for Wind Measurements in Upper Troposphere and Stratosphere. *Remote Sens.* **2022**, *14*, 5556. [CrossRef]
26. Chen, J.; Xie, C.; Zhao, M.; Ji, J.; Wang, B.; Xing, K. Research on the Performance of an Active Rotating Tropospheric and Stratospheric Doppler Wind Lidar Transmitter and Receiver. *Remote Sens.* **2023**, *15*, 952. [CrossRef]

photonics

MDPI

Article

Research on the Correction Algorithm for Ozone Inversion in Differential Absorption Lidar

Leyong Li [1,2], Chenbo Xie [1,*], Jie Ji [1,2] and Kunming Xing [1]

[1] State Key Laboratory of Laser Interaction with Matter, Anhui Institute of Optics and Fine Mechanics, HFIPS, Chinese Academy of Sciences, Hefei 230031, China; lly9977@mail.ustc.edu.cn (L.L.); jijie3496@mail.ustc.edu.cn (J.J.); kunmingx@mail.ustc.edu.cn (K.X.)

[2] Science Island Branch of Graduate School, University of Science and Technology of China, Hefei 230026, China

* Correspondence: cbxie@aiofm.ac.cn

Abstract: Due to the complex and variable nature of the atmospheric conditions, traditional multi-wavelength differential absorption lidar (DIAL) methods often suffer from significant errors when inverting ozone concentrations. As the detection range increases, there is a higher demand for Signal to Noise Ratio (SNR) in lidar signals. Based on this, the paper discusses the impact of different atmospheric factors on the accuracy of ozone concentration inversion. It also compares the advantages and disadvantages of the two-wavelength differential method and the three-wavelength dual-differential method under both noisy and noise-free conditions. Firstly, the errors caused by air molecular extinction, aerosol extinction, and backscatter terms in the inversion using the two-wavelength differential method were simulated. Secondly, the corrected inversion errors were obtained through direct correction and the introduction of a three-wavelength dual differential correction. Finally, addressing the issue of insufficient SNR in practical inversions, the inversion errors of the two correction methods were simulated by constructing lidar parameters and incorporating appropriate noise. The results indicate that the traditional two-wavelength differential algorithm is significantly affected by aerosols, making it more sensitive to aerosol concentration and structural changes. On the other hand, the three-wavelength dual differential algorithm requires a higher SNR in lidar signals. Therefore, we propose a novel strategy for inverting atmospheric ozone concentration, which prioritizes the use of the three-wavelength dual-differential method in regions with high SNR and high aerosol concentration. Conversely, the direct correction method utilizing the two-wavelength differential approach is used. This approach holds the potential for high-precision ozone concentration profile inversion under different atmospheric conditions.

Keywords: differential absorption lidar; ozone inversion; aerosol; signal to noise ratio; inversion strategy

1. Introduction

Ozone is an important component of the Earth's atmosphere, playing a crucial role in atmospheric radiation. Stratospheric ozone effectively absorbs ultraviolet radiation harmful to biological life; meanwhile, tropospheric ozone acts as a significant atmospheric pollutant. High concentrations of ground-level ozone can lead to photochemical smog. Prolonged exposure to high levels of ozone can cause damage to various organs in the human body [1–5]. Ozone also affects the growth and development of plants [6–10] and can cause corrosion to building materials [2,11].

Since the observation of the ozone layer hole over Antarctica in 1986, the distribution and changes in ozone have received widespread attention [3,12]. With the development of industrialization, pollution from tropospheric ozone has been deepening annually [13,14], making ozone concentration a very important key factor in assessing the atmospheric environment [15]. Ozone measurement is the basis for studying ozone changes. Differential

Absorption Lidar (DIAL) technology, with its advantages in high temporal resolution, spatial resolution, and measurement accuracy, has become an important means of observing ozone [16–20].

The Differential Absorption Lidar (DIAL) technology was first proposed by Professor Schotland of the University of Michigan in 1964 and applied to the detection of atmospheric water vapor content [20]. Subsequently, it has been extended to the concentration detection of trace gases such as ozone, nitrogen dioxide, sulfur dioxide, etc. [21–23].

The DIAL technology measures ozone concentration by utilizing the differential absorption of laser wavelengths by ozone. The lidar emits two laser beams of different wavelengths into the atmosphere and receives the return signals that have interacted with the atmosphere. It is assumed that the differences in the return signals of these two laser beams are caused by the differential absorption by ozone, thus allowing the determination of ozone concentration through a differential algorithm. However, due to the presence of atmospheric aerosols, aerosols often also have different effects on these two wavelengths.

When the difference in wavelengths between the two laser beams is not sufficiently small, the difference in the absorption cross-section of ozone for these two wavelengths is not significant enough, or if there is a substantial and uneven distribution of aerosols along the laser path, it becomes necessary to correct the differential absorption formula.

To mitigate the impacts of aerosol extinction and backscattering, scholars worldwide have undertaken considerable work. A portion of this work is dedicated to theoretical corrections, estimating aerosol extinction and backscatter characteristics through lidar signals at other wavelengths or other means. Browell and colleagues discussed the errors caused by differences in atmospheric wavelengths and aerosol inhomogeneity and corrected the ozone inversion results [24]. Steinbrecht and colleagues quantified the impact of additional aerosols on differential absorption measurements and made corrections [25]. Papayannis used a multi-wavelength lidar system as a reference for correcting ozone measurements [26]. Due to the inability to accurately obtain aerosol parameters, errors still exist in these methods.

Another line of work does not concern itself with the state of aerosols along the laser path but instead aims to reduce the impacts of aerosols through experimental methods. McGee and others proposed the Raman-DIAL method, which uses the Raman backscatter from nitrogen molecules at two wavelengths for differential measurement, eliminating the impact of aerosol backscatter [27]. However, the impact of aerosol extinction still exists, and since the Raman scattering cross-section is two orders of magnitude smaller than Rayleigh and Mie scattering, this method has a higher requirement for the signal-to-noise ratio. Wang Zhen and colleagues introduced the three-wavelength dual-differential method (Dual-DIAL), incorporating a third wavelength. By performing a double differential operation across three wavelengths, this method reduces the impact of aerosol extinction and backscatter [28,29]. A limitation is the smaller difference in extinction cross-section, leading to reduced sensitivity and spatial resolution in measuring air molecules and potentially resulting in insufficient signal-to-noise ratio.

This article mainly discusses the impact of various atmospheric factors on the accuracy of ozone inversion. By combining the signal-to-noise ratio issues encountered in actual inversions, it simulates and compares the errors between direct correction with DIAL and the three-wavelength dual-differential method. A multi-wavelength differential absorption correction algorithm is proposed. By integrating the above, adjusting strategies under different atmospheric factors and lidar parameters can yield better results, showing potential application value for the inversion of high-precision ozone concentration profiles.

2. Inversion Principle and Simulation Approach

2.1. Inversion Principle

Differential Absorption Lidar emits two lasers with different wavelengths into the atmosphere, one as the measurement wavelength (λ_{on}) and the other as the reference wavelength (λ_{off}). When these two wavelengths are sufficiently close, it can be assumed that components in the air other than ozone have the same effect on both wavelengths. Therefore, the difference in the return signals is considered to be entirely caused by the differential absorption of ozone. By comparing the return signals of the two wavelengths, the concentration of ozone can be calculated. Due to the unique ultraviolet absorption cross-section of ozone, without considering the absorption by other gases, the echo signal expression for differential absorption lidar is as follows [20]:

$$P(\lambda, z) = C(\lambda, z)\frac{\beta(\lambda, z)}{z^2}exp\left[-2\int_{z_0}^{z}\alpha(z) + N(z)\delta(\lambda, z)dz\right] \tag{1}$$

In the above formula, $P(\lambda, z)$ represents the return signal at height z for wavelength λ, $C(\lambda, z)$ is the lidar constant determined by the lidar's parameters, $\beta(\lambda i, z)$ represents the atmospheric backscatter signal, $\alpha(\lambda i, z)$ represents the atmospheric extinction, which includes air molecules and aerosols, $N(z)$ represents the ozone concentration, and $\delta(\lambda, z)$ represents the absorption cross-section of ozone molecules at wavelength λ.

$$C = cE_0(\lambda)\Omega(\lambda, z)A_r T_t(\lambda)T_r(\lambda) \tag{2}$$

The expression for the lidar constant includes the following elements: c is the speed of light, $E_0(\lambda)$ is the single pulse energy of the laser, $\Omega(\lambda, z)$ is the lidar's geometric factor, A_r is the area of the receiving telescope, $T_t(\lambda)$ and $T_r(\lambda)$ are the total transmittance of the transmitting and receiving optical units at wavelength λ, respectively.

The ozone concentration is calculated from the echo signal equations of two wavelengths as follows:

$$N(z) = \frac{-1}{2\Delta\sigma}\frac{d}{dz}\left[ln\frac{P(\lambda_{on}, z)}{P\left(\lambda_{off}, z\right)}\right] + B - E_a - E_m \tag{3}$$

where $\Delta\sigma$ is the difference in the absorption cross-section of ozone for the two wavelengths, $P(\lambda_{on}, z)$ and $P\left(\lambda_{off}, z\right)$ are the lidar signals for the two wavelengths. B is the error caused by the atmospheric backscatter, including atmospheric air molecules and aerosols. E_a is the extinction error due to aerosols, and E_m is the extinction error due to air molecules, collectively referred to as atmospheric interference terms. The specific expressions for the three terms are as follows:

$$B = \frac{1}{2\Delta\sigma}\frac{d}{dz}\left[ln\frac{\beta(\lambda_{on}, z)}{\beta\left(\lambda_{off}, z\right)}\right] \tag{4}$$

$$E_a = \frac{1}{\Delta\sigma}\left(\alpha_a(\lambda_{on}, z) - \alpha_a\left(\lambda_{off}, z\right)\right) \tag{5}$$

$$E_m = \frac{1}{\Delta\sigma}\left(\alpha_m(\lambda_{on}, z) - \alpha_m\left(\lambda_{off}, z\right)\right) \tag{6}$$

where $\alpha_m(\lambda_{on}, z)$ and $\alpha_m\left(\lambda_{off}, z\right)$ are the extinction coefficients of air molecules at wavelengths λ_{on} and λ_{off}, respectively.$\alpha_a(\lambda_{on}, z)$ and $\alpha_a(\lambda_{on}, z)$ are the extinction coefficients of aerosols at wavelengths λ_{on} and λ_{off}, respectively. $\beta(\lambda_{on}, z)$ and $\beta\left(\lambda_{off}, z\right)$ represent the backscatter coefficients of the atmosphere at wavelengths λ_{on} and λ_{off}, respectively.

2.1.1. Direct Correction

To reduce the error in the inversion of ozone, the common practice is to obtain the extinction coefficients of aerosols and air molecules, as well as the backscatter coefficients, and then substitute these values into Formulas (4)–(6).

The extinction and backscatter coefficients of air molecules can be calculated using the US Standard Atmosphere Model (USSA-76), while the relevant parameters for aerosols can usually be obtained through the Fernald method. The Fernald method calculates the aerosol extinction coefficient as follows [30]:

$$\alpha_a(z) = -\frac{S_a}{S_m}\alpha_m(z) + \frac{X(z)\exp\left[2\left(\frac{S_a}{S_m}-1\right)\int_{z_0}^{z_c}\alpha_m(z)dz\right]}{\frac{X(z_c)}{\alpha_m(z_c)+\frac{S_a}{S_m}\alpha_m(z_c)} + 2\int_{z_0}^{z} X(z')\exp\left[2\left(\frac{S_a}{S_m}-1\right)\int_{z_0}^{z_c}\alpha_m(z)dz\right]dz'} \tag{7}$$

where $X(z) = P(z)z^2$ is the distance squared signal and z_c is the correction layer height at which the aerosol extinction coefficient is assumed to be known.

The Fernald method assumes $S_m = \alpha_m(z)/\beta_m(z)$ and $S_a = \alpha_a(z)/\beta_a(z)$, where S_m is the ratio of molecular extinction to backscatter, and this value can be solved using Rayleigh scattering theory as $8\pi/3$ [30]. S_a is set as the ratio of aerosol extinction to backscatter; it is a constant that does not change with altitude, and its value typically varies between 10 to 100 sr. A commonly used value is 50 sr.

The aerosol Ångström exponent δ describes the relationship between the scattering and absorption properties of aerosols at different wavelengths. This value typically varies between 0 and 2. The aerosol extinction coefficients at different wavelengths can be expressed as follows [31]:

$$\delta = ln(\frac{\alpha(\lambda_1)}{\alpha(\lambda_2)})/ln(\frac{\lambda_2}{\lambda_1}) \tag{8}$$

By obtaining the aerosol extinction coefficient at wavelength λ_1 through the Fernald method and converting it to the detection wavelength λ_{on} and the reference wavelength λ_{off} using the Ångström exponent δ, direct correction of the differential absorption algorithm is accomplished.

2.1.2. Dual-DIAL Correction

A three-wavelength differential absorption lidar emits three laser beams of different wavelengths into the atmosphere, corresponding to strong, medium, and weak (or non-) absorption by the gas under measurement. By combining the three wavelengths in pairs, two sets of differential absorption equations can be derived. The three-wavelength differential absorption formula aims to negate atmospheric interference by combining these two differential absorption formulas, thereby eliminating or reducing the interference caused by aerosols. The ozone concentration obtained through the double differential of three wavelengths is as follows [28]:

$$N(z) = \frac{1}{2\Delta\sigma}\frac{d}{dz}\left[-ln\frac{P(\lambda_{1on},z)}{P\left(\lambda_{1off},z\right)} + Cln\frac{P(\lambda_{2on},z)}{P\left(\lambda_{2off},z\right)}\right] \tag{9}$$

In this context, where $\lambda_{1off} = \lambda_{2on}$, and C is a constant, typically chosen as $C = \left(\lambda_{1on} - \lambda_{1off}\right)/\left(\lambda_{2on} - \lambda_{2off}\right)$.

2.2. Simulation Approach

The accuracy of ozone concentration monitoring is constrained by atmospheric environmental factors, as well as the parameters of the radar itself. The inversion of ozone concentration involves complex atmospheric processes and physical mechanisms. It is necessary to construct a detailed simulation environment, precisely setting the variables and conditions in the atmosphere to ensure the repeatability and reliability of the results. While controlling and understanding these complex processes, it is also essential to reveal the intrinsic relationship between algorithm performance and atmospheric parameters and to evaluate the performance of the algorithm under different atmospheric conditions.

This study establishes a simulated atmospheric environment and simulates the interaction process between the lidar and the atmosphere under various aerosol parameters (aerosol backscatter ratio S_a and Ångström exponent δ), obtaining lidar echo signals at different wavelengths. By combining Equations (1)–(3), the inverted ozone concentration is derived, followed by an error analysis.

Figure 1 illustrates the basic atmospheric environment established for the simulations in this section, where Figure 1a displays the aerosol extinction coefficient profile at 316 nm as set; Figure 1b shows the extinction coefficient profile for air molecules at 316 nm; and Figure 1c outlines the set atmospheric ozone concentration profile.

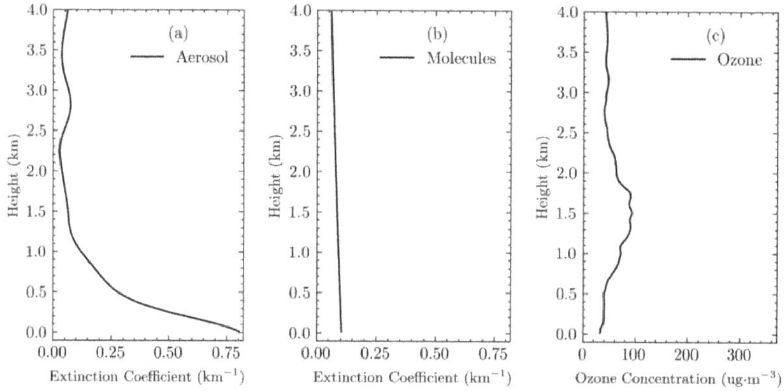

Figure 1. The Established Basic Atmospheric Environment. (**a**) Aerosol Extinction Coefficient. (**b**) Air Molecular Extinction Coefficient. (**c**) Ozone Concentration.

The aerosol extinction coefficients and ozone concentrations set in this section are slightly modified based on actual measurement data, whereas the extinction coefficients for atmospheric air molecules are provided by the model (USSA-76). It is generally considered that the atmospheric air molecule model closely fits the real situation, and the errors caused by atmospheric air molecules can be effectively eliminated. The setting of the atmospheric aerosol extinction coefficient highlights the higher concentration of aerosols near the ground surface, with a rapid decrease in the aerosol extinction coefficient with increasing altitude. In this profile, the change in the aerosol extinction coefficient is not completely smooth; some abrupt structural changes are set at 0.5 km, 1.2 km, and 2.3 km to better understand the composition and magnitude of the ozone inversion error at the points of the aerosol structure change.

3. Simulation Results

This section first simulates the various errors of atmospheric disturbance terms in the differential absorption algorithm, analyzing their causes and proportions. Secondly, it analyzes the sensitivity of the two correction methods to aerosol parameters. Finally, considering the inevitable noise issues in actual detection, it compares the ozone inversion accuracy of the two correction methods under different atmospheric conditions. In the simulation process of this paper, the wavelength pair used is 266–316 nm. During the correction process using the triple-wavelength double-differential method, the signal at 289 nm is added.

3.1. Error Analysis of Atmospheric Interference Terms

When retrieving ozone concentration using the dual-wavelength differential absorption algorithm, errors related to the extinction terms of aerosols, atmospheric air molecules, and backscattering terms occur, collectively referred to as errors of atmospheric interference terms. Initially, the causes and proportions of these three errors are analyzed.

Figure 2 presents the error of atmospheric interference terms under different aerosol backscatter ratio (S_a) and Ångström exponent (δ), where the aerosol backscatter ratio is denoted as SA and the Ångström exponent as AE in the figure. Since the extinction by atmospheric air molecules is calculated using model data, the error associated with the atmospheric air molecule term remains constant. The error in the aerosol extinction term is directly related to aerosol concentration. When the height is greater than 2 km, where the aerosol concentration is low, the error in the aerosol extinction term is minimal. Compared to the aerosol backscatter ratio, the error in the aerosol extinction term is more sensitive to the Ångström exponent. The closer the Ångström exponent is to 0, meaning the closer the aerosol extinction coefficients are between different wavelengths, the smaller the error in the aerosol extinction term becomes. Overall, the error in the backscattering term also decreases with a decrease in aerosol concentration. However, in areas where aerosol structure changes, such as the 2–3 km region, the error in the backscattering term increases. In regions with higher aerosol concentration, the error in the aerosol extinction term dominates; when the aerosol concentration decreases and is at a point of aerosol structure change, the error in the backscattering term increases and becomes dominant.

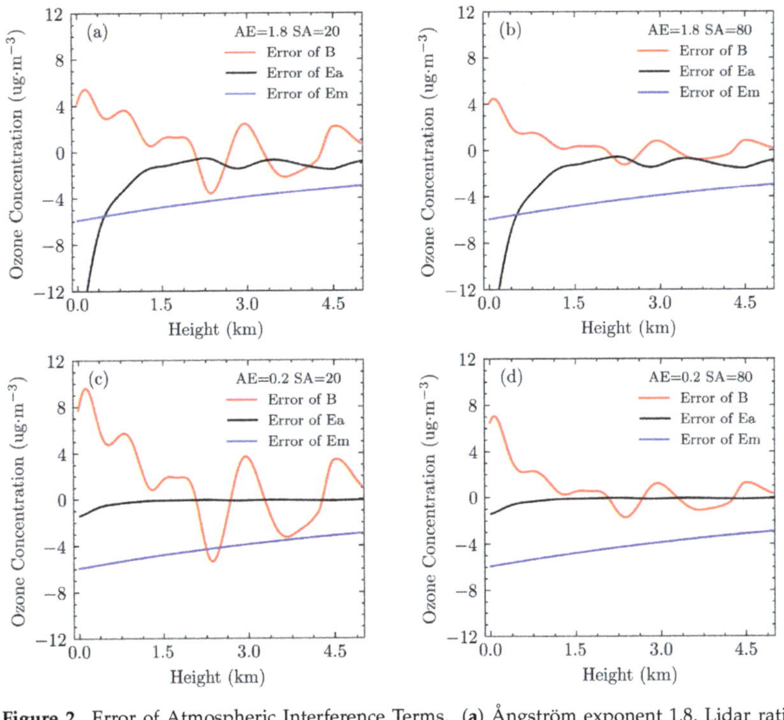

Figure 2. Error of Atmospheric Interference Terms. (**a**) Ångström exponent 1.8, Lidar ratio 20 (**b**) Ångström exponent 1.8,Lidar ratio 80 (**c**) Ångström exponent 0.2,Lidar ratio 20 (**d**) Ångström exponent 0.2, Lidar ratio 80.

3.2. *Sensitivity Analysis of Aerosol Parameters*

This section will discuss the sensitivity of the ozone inversion errors obtained by the two ozone inversion correction methods described in Section 2.1 to aerosol parameters.

3.2.1. Sensitivity Analysis of Direct Correction

The variation of ozone concentration error with aerosol backscatter ratio and Ångström exponent at different heights is illustrated in Figure 3. The Ångström exponent (AE) varies between 0 and 2, and the lidar ratio (SA) changes between 10 and 90. It is observed that at different heights, there is a certain pattern in the error of ozone inversion. Negative maximum inversion errors occur when the Ångström exponent and lidar ratio are close to 0 and 10, respectively; positive maximum inversion errors appear when they are close to 2 and 90, respectively. Furthermore, it is noteworthy that the direct corrected dual-wavelength algorithm exhibits higher sensitivity to variations in the Ångström exponent.

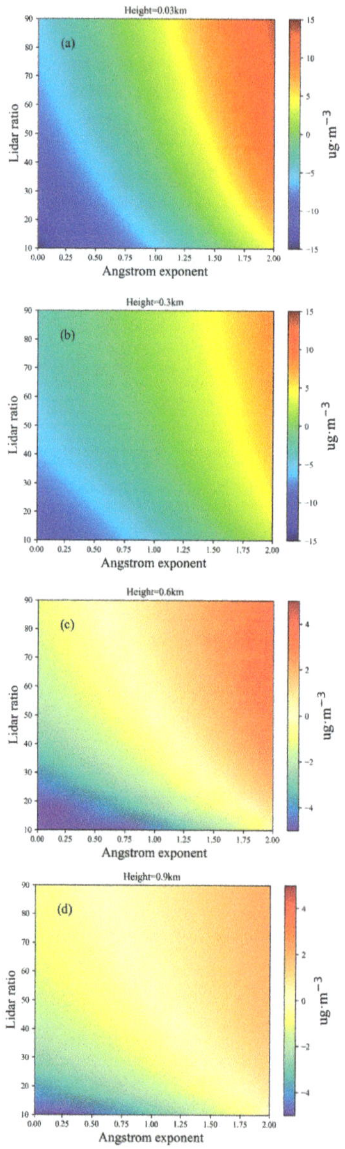

Figure 3. The relationship between ozone inversion error after direct correction and aerosol backscatter ratio and Ångström exponent. (**a**) The height of 0.03 km. (**b**) The height of 0.3 km. (**c**) The height of 0.6 km. (**d**) The height of 0.9 km.

As depicted in Figure 3a,b, the aerosol extinction coefficient at 316 nm is measured as 0.76 km^{-1} and 0.43 km^{-1} at heights of 0.03 km and 0.3 km, respectively, wherein the direct correction dual-wavelength method may exhibit an ozone concentration inversion error exceeding 15 μg/m^3. As illustrated in Figure 3c,d, the aerosol extinction coefficient at 316 nm is 0.21 km^{-1} and 0.18 km^{-1} at heights of 0.03 km and 0.3 km, respectively, with the ozone inversion error being within 5 μg/m^3.

By selecting the Ångström exponent values of 0.2 and 1.8 and lidar ratios of 20 and 80, four sets of ozone concentration inversion profiles and their error curves are obtained. In Figure 4a, the black line represents the set ozone concentration, and the dash-dot lines represent several inversion profiles from the direct correction inversion algorithm. Figure 4b shows their error curves, with the four error curves exhibiting similar structures. Below 1 km, where the aerosol extinction coefficient is larger, the inversion error is greater. The farther the parameters of the error curves deviate from the correction parameters (AE = 1, SA = 50), the greater the ozone inversion error. As altitude increases and aerosol concentration decreases, the inversion results tend to converge.

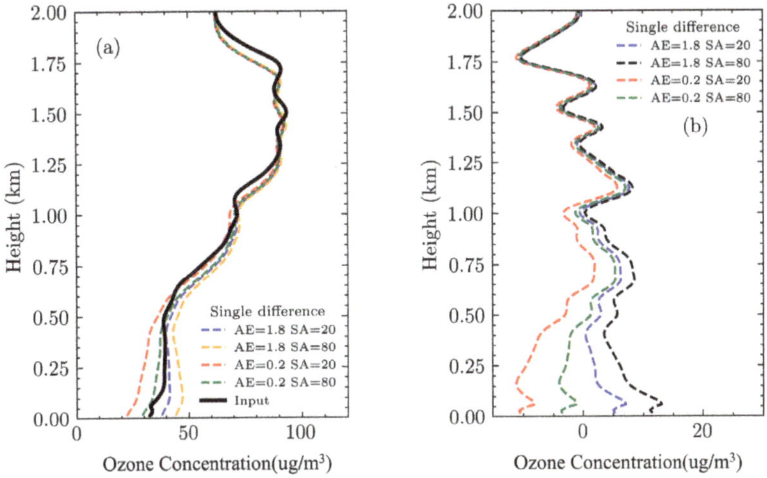

Figure 4. Ozone inversion profile and error profile using the direct correction differential algorithm. (**a**) Ozone inversion profile (**b**) error profile.

3.2.2. Sensitivity Analysis of Dual-DIAL Correction

The triple-wavelength double-differential correction, through the processing of two sets of differential results, has mitigated some of the effects caused by aerosols. This subsection analyzes the variation in inversion error of this correction method with the aerosol backscatter ratio and Ångström exponent and presents the ozone inversion and error profiles derived from the triple-wavelength algorithm.

As illustrated in Figure 5, while varying the aerosol backscatter ratio (SA) and Ångström exponent (AE) to simulate the variation in real aerosol parameters, the inversion error from the triple-wavelength double-differential correction does not exhibit a clear pattern with respect to the Ångström exponent and lidar ratio, unlike the direct correction method. The triple-wavelength double-differential correction effectively reduces the influence of aerosols. At an altitude of 0.03 km with an aerosol extinction coefficient (at 316 nm) of 0.76 km^{-1}, the inversion error from the triple-wavelength double-differential correction is within 4 μg/m^3, whereas the direct correction error can reach 15–18 μg/m^3. At an altitude of 0.3 km, with an aerosol extinction coefficient (at 316 nm) of 0.43 km^{-1}, the inversion error for ozone is already within 3 μg/m^3. At altitudes of 0.6 km and 0.9 km, the error from the triple-wavelength double-differential correction is also smaller than that from the direct correction method.

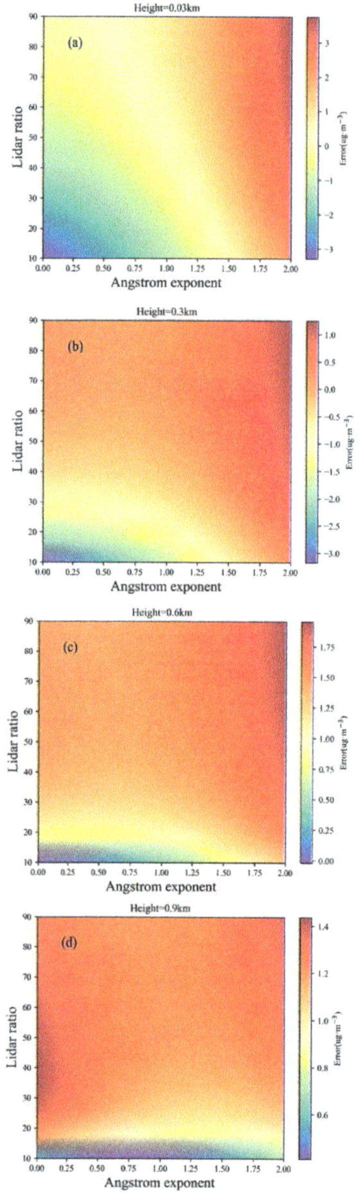

Figure 5. The relationship between ozone inversion error after Dual-DIAL correction and aerosol backscatter ratio and Ångström exponent. (**a**) The height of 0.03 km. (**b**) The height of 0.3 km. (**c**) The height of 0.6 km. (**d**) The height of 0.9 km.

Selecting aerosol backscatter ratios of 20 and 80 and Ångström exponents of 0.2 and 1.8, four sets of ozone concentration inversion profiles and error curves are obtained through the triple-wavelength double-differential correction. In Figure 6a, the black line represents the preset ozone concentration, while the dash–dot lines are several inversion profiles from the triple-wavelength double-differential correction. Figure 6b displays the corresponding error curves. The four error curves have similar structures, but their relationship with aerosol parameters is not clear.

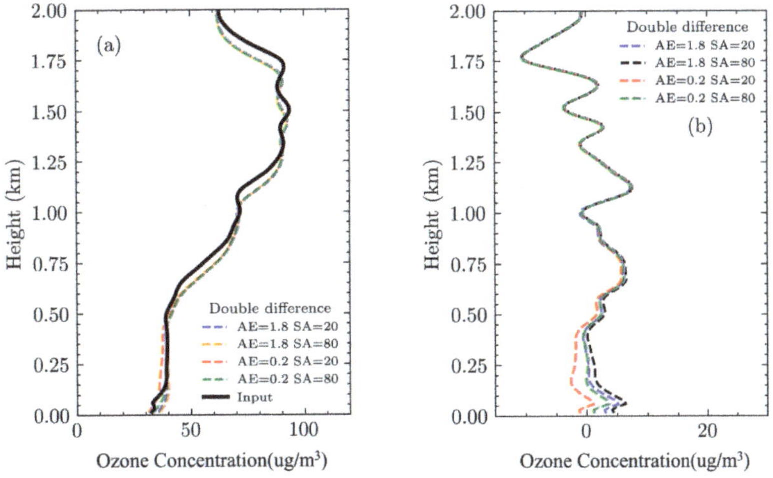

Figure 6. Ozone inversion profile and error profile using the Dual-DIAL correction differential algorithm. (**a**) Ozone inversion profile (**b**) error profile.

3.3. Comparison of Algorithm Performance under Noise Conditions

In the actual detection process, the interference of noise is inevitable. Noise affects the performance of lidar and the accuracy of measurement results, directly impacting measurement precision and system performance. Incorporating noise factors into simulation design can enhance the accuracy and reliability of the simulation, providing a better understanding of the challenges and limitations the system may face in actual operation and making the simulation results more closely aligned with real-world conditions. Evaluating the impact of noise on system performance allows for a more accurate prediction of the system's behavior under real conditions. Studying the impact of noise on signal processing aims to improve the quality of the signal and the overall performance of the system.

In differential absorption lidar systems, noise sources can be divided into background light noise $N_b(\lambda, z)$ and detector noise $Nd(\lambda, z)$. The background light noise, originating from solar radiation or other artificial light sources, can enter the detector through the optical components of the receiving system and mix with the signal light, leading to noise. Detector noise includes thermal noise, shot noise, dark current noise, etc., all of which are generated internally by the detector. Among these, shot noise, which is related to the number of photons detected, is a manifestation of quantum noise. These two types of noise can be expressed by the following Equations:

$$N_b(\lambda) = \frac{\eta(\lambda)\lambda}{hc} P_b(\lambda)\pi\frac{\theta}{2}\Delta\lambda(\lambda)A_r T_r \Delta t \quad (10)$$

$$N_d(\lambda) = C_{PS}(\lambda)\Delta t \quad (11)$$

$P_b(\lambda)$ is defined as the sky background radiance at wavelength λ. At night, this value is close to 0; thus, operating a DIAL system at night can significantly reduce the impact of background light noise. θ is the receiving field of view angle of the receiving telescope; Δλ(λ) is the half-width of the spectral device at wavelength λ; T_r is the transmittance of the receiving optical unit at wavelength λ, and $C_{PS}(\lambda)$ is the dark count of the detector. The signal-to-noise ratio SNR(λ,z) of the atmospheric backscatter signal from the differential absorption lidar can be calculated using the formula below:

$$SNR(\lambda, Z) = \frac{N_s(\lambda, Z)}{\sqrt{N_s(\lambda, Z) + 2(N_b(\lambda, Z) + N_d(\lambda, Z))}}\sqrt{M} \tag{12}$$

where M is the number of pulses.

Combining the parameters from the table above simulates the atmospheric transmission process in the air. Introduce noise into the lidar equation. Calculate the echo signals at 266 nm, 289 nm, and 316 nm.

The signal-to-noise ratio (SNR) for the three wavelengths is obtained by substituting the echo signals into the SNR formula, as shown in Figure 7. The detection altitudes for 266 nm, 289 nm, and 316 nm, where the SNR is 10, are respectively 1.725 km, 3.18 km, and 4.09 km. In the ultraviolet spectrum, as the wavelength decreases, the atmospheric backscatter effect intensifies, leading to a faster attenuation of signal strength. Therefore, there are significant differences in the effective detection distances among the signals at the three different wavelengths. Within these effective detection ranges, the variation in the echo signals across the three wavelengths can reach up to five orders of magnitude.

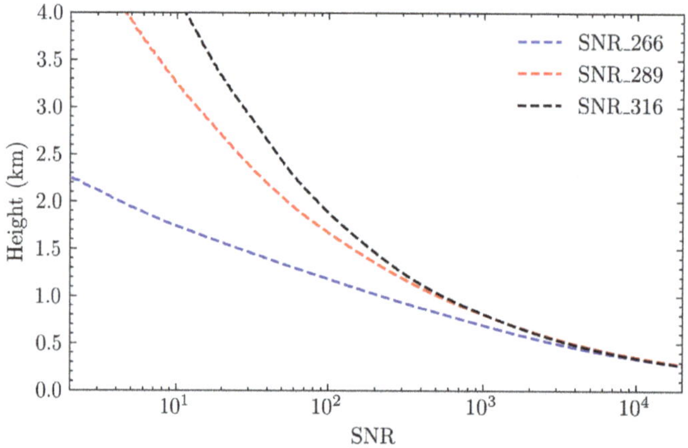

Figure 7. Signal-to-noise ratio for the three wavelengths.

Under the lidar system parameters listed in Table 1, by varying the aerosol backscatter ratio and Ångström exponent, ozone concentration profiles were retrieved through both direct correction and Dual-DIAL correction, as shown in Figure 8. Figure 8a presents the results from direct correction, and Figure 8b from the Dual-DIAL correction. The inversion spatial resolution was set at 60 m. In high signal-to-noise ratio (SNR) areas (such as below 1 km), the ozone profiles are essentially consistent with the ideal state (in the absence of noise); as the SNR of the wavelength signal decreases, fluctuations appear in the inverted ozone concentration.

Table 1. Simulated Lidar Parameters for Calculation.

System Parameter	Value
Laser wavelength	266 nm, 289 nm, 316 nm
Pulse energy	1 mJ
Pulse number	700
Emitting optical transmittance	0.3
Detection Quantum Efficiency	0.2
Receiver Diameter	300 mm
Field of view	1 mrad
Filter Bandwidth	1 nm
Dark counts	400
Spatial resolution	7.5 m

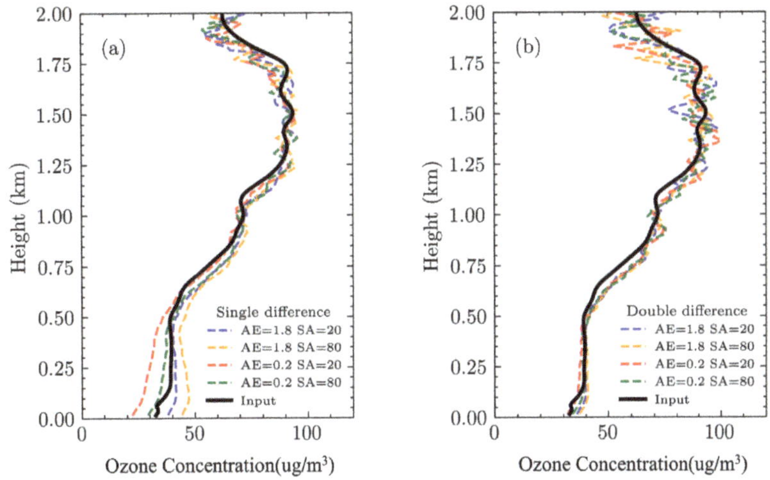

Figure 8. Comparison of inversion errors between direct correction and Dual-DIAL Correction under the presence of noise. (**a**) Single difference. (**b**) Double difference.

The results depicted in Figure 8 demonstrate the efficacy of employing a Dual-DIAL correction in mitigating aerosol-induced influences; however, it necessitates a higher signal-to-noise ratio, as evident from the findings.

4. Error Caused by Noise

The error in Figure 8 can be divided into two parts: one part is the error caused by the interference of aerosols, and the other part is the error generated by system noise. By comparing with the ozone inversion profile under ideal conditions, the magnitude of errors caused solely by noise for both algorithms can be obtained, as shown in Figure 9.

Figure 9a shows the error caused by noise in direct correction, while Figure 9b shows the error caused by noise in Dual-DIAL correction. It is observed that when the error reaches 5 $\mu g/m^3$, the inversion heights for direct correction and Dual-DIAL correction are 1.3 km and 0.8 km, respectively. When the inversion height reaches 2 km, the errors caused by noise in direct correction and Dual-DIAL correction are 8 $\mu g/m^3$ and 15 $\mu g/m^3$, respectively.

Combining Figures 8 and 9, it can be seen that in the high-concentration aerosol areas near the ground, the Dual-DIAL correction performs better, as it can reduce most of the aerosol interference. As the altitude increases and the lidar signal's signal-to-noise ratio becomes insufficient, the direct correction method is less affected by noise and yields better inversion results. From this, we can conclude the following strategy for actual detection: use the Dual-DIAL correction in areas of high aerosol concentration and the direct correction method in areas of low aerosol concentration. This inversion strategy can effectively improve the accuracy of ozone inversion and the detection distance of ozone.

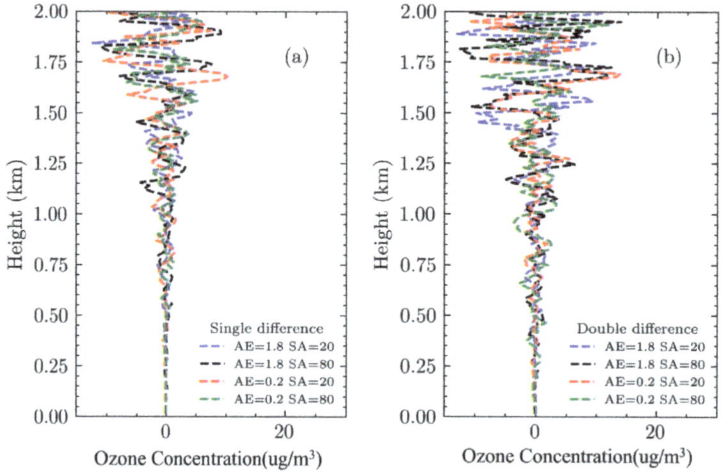

Figure 9. Error Caused by Noise. (**a**) Single difference. (**b**) Double difference.

5. Discussion

Studying the distribution of atmospheric ozone concentration in the troposphere is of significant importance for understanding atmospheric chemistry, assessing environmental impacts, and studying climate change. Compared to other detection methods, differential absorption lidar (DIAL) exhibits unique advantages in detecting ozone concentration distribution, including high-resolution vertical profiles, high temporal and spatial resolution, multi-parameter measurement capabilities, high sensitivity, and long-range detection. These advantages make lidar a powerful tool for researching and monitoring ozone distribution.

From a theoretical algorithm perspective, the main challenge of using DIAL for detecting ozone concentration distribution lies in the interference from atmospheric aerosols. In traditional inversion algorithms, the wavelength dependence of aerosol extinction on laser radiation is unknown, and it is mistakenly calculated as ozone concentration. Particularly in the region from near the surface to the troposphere, on the one hand, the ozone concentration in the troposphere is relatively low compared to the stratosphere, and on the other hand, anthropogenic activities and industrial production contribute to higher aerosol content in the atmosphere in this region.

In order to mitigate the impact of aerosols on the DIAL algorithm, the current mainstream research directions can be divided into two categories. One approach does not focus on the optical properties of aerosols along the laser path but instead aims to compensate for or reduce atmospheric interference through experimental methods.

For example, McGee utilized the backscattered signal of the receiving laser on nitrogen molecules to obtain ozone concentration profiles under high aerosol concentrations. It is widely recognized that this method effectively eliminates the influence of aerosol backscattering error. However, errors in aerosol extinction still persist [27]. Wang et al. proposed a three-wavelength dual differential absorption lidar technique and used a three-wavelength dual-wavelength lidar system to detect ozone in the stratospheric region after a volcanic eruption [29]. Su et al. conducted monitoring of the lower atmospheric vertical profile in the urban area of Hangzhou using a three-wavelength dual differential absorption algorithm and compared the results with measurements obtained from radiosonde balloons [32]. Yang et al. simulated the production and diffusion process of ozone, evaluated the performance of ozone concentration inversion using a multi-wavelength DIAL system, and conducted verification in the Chengdu region of China [33].

Another approach involves detecting the optical properties of aerosols along the laser path through other means to correct the inverted ozone concentration. For example, Ma et al. analyzed the atmospheric backscattering signals at three wavelengths, 266 nm, 289 nm, and 316 nm, to obtain ozone concentrations. They then used the backscattered signal at 532 nm to correct for the aerosol optical effects [34]. Lei et al. studied the impact of aerosols emitted from wildfires on ozone concentration inversion using Differential Absorption Lidar. They obtained aerosol extinction coefficients at 532 nm and 292 nm through different methods and determined the Ångström exponent of the aerosols to correct for their influence [35]. Kuang et al. utilized a High Spectral Resolution Lidar (HSRL) to retrieve the aerosol extinction coefficients at 532 nm and 340 nm. They combined the backscattered signals from the DIAL to invert the aerosol extinction coefficient profiles and Ångström exponent and subsequently corrected the ozone concentration [36].

These two research directions provide important references for accurately inverting ozone concentrations in the atmosphere. However, they also have some unresolved limitations. Correcting the ozone concentration using experimental methods can theoretically only remove a portion of the aerosol influence and impose higher requirements on the signal-to-noise ratio(SNR) of the lidar signals. On the other hand, the other research direction is evolving towards joint inversion of aerosol and ozone concentrations. However, it faces the challenge that the state of aerosols cannot be precisely determined, leading to uncertainties in correcting the ozone concentration.

Based on the aforementioned background, this study delves into the mechanisms of aerosol impact on ozone concentration inversion. By combining predefined atmospheric conditions and lidar parameters, the inversion effects of different correction methods are simulated and investigated. A novel strategy for differential absorption ozone concentration inversion is proposed. This strategy utilizes a three-wavelength dual differential algorithm for correction under conditions of high aerosol concentrations and high SNR. Conversely, a direct correction method is employed for correction under other conditions. Compared to traditional methods, this strategy offers higher inversion accuracy and longer detection distances.

The work presented in this paper is not specific to any particular lidar system but focuses on analyzing and studying the constraints of the ozone inversion algorithms themselves. The constraints depend on the actual atmospheric conditions and the SNR of the lidar signals. Therefore, it is not necessary to provide specific lidar parameters or quantified boundary conditions.

6. Conclusions

This paper conducts a simulation study on the inversion of ozone concentration using the differential absorption lidar algorithm. Initially, under ideal conditions, the atmospheric interference terms of the differential absorption algorithm are simulated and calculated, analyzing their causes and proportions. Subsequently, a sensitivity analysis of aerosol parameters is performed for both direct correction and triple-wavelength double-differential correction. Finally, considering the inevitable noise issues in actual detection, the inversion accuracy of these two correction methods under different atmospheric conditions is compared. A better inversion strategy is proposed.

The simulation results indicate the following:

For the atmospheric interference terms of the differential absorption algorithm, the error caused by aerosol extinction dominates under conditions of higher aerosol concentration and stable aerosol structural changes. In contrast, the error caused by backscattering dominates under conditions of lower aerosol concentration and changing aerosol structure.

Compared to the aerosol backscatter ratio, the ozone concentration inversion error obtained through direct correction is more sensitive to the aerosol's Ångström exponent. However, the sensitivity of the ozone concentration inversion error from triple-wavelength double-differential correction to these two parameters does not exhibit a clear pattern. In the absence of noise, triple-wavelength double-differential correction can better reduce the impact of aerosols on ozone concentration inversion.

Considering the inevitable noise issues in actual detection and adjusting aerosol parameters, a simulation analysis of the ozone concentration inversion error for both correction methods was conducted. In areas of high SNR and high aerosol concentration, the inversion accuracy of both correction methods is close to the ideal state, with the triple-wavelength double-differential correction achieving higher ozone concentration inversion accuracy. With increasing height, both SNR and aerosol concentration decrease, leading to dominant errors induced by noise and consequently resulting in improved inversion accuracy when employing the direct correction method.

In summary, the triple-wavelength double-differential correction method can be used in areas with high SNR and high aerosol concentration, while the direct correction method is suitable for areas with low SNR and low aerosol concentration. This strategy for ozone concentration inversion enables enhanced accuracy and an extended range of inversion.

Author Contributions: Supervision, conceptualization, C.X.; resources, K.X.; software, L.L.; methodology, J.J.; numerical simulation, writing—original draft preparation, L.L.; writing—review and editing, C.X. All authors have read and agreed to the published version of the manuscript.

Funding: This work was supported in part by the Strategic Priority Research Program of the Chinese Academy of Sciences (XDA17040524), Anhui Province science and technology major project (201903c08020013).

Institutional Review Board Statement: Not applicable.

Informed Consent Statement: Not applicable.

Data Availability Statement: Data are contained in this article.

Acknowledgments: The authors thank Xie Chenbo, Anhui Institute of Optics and Fine Mechanics, Hefei Institute of Physical Sciences, Chinese Academy of Sciences for his theoretical guidance and detailed discussion.

Conflicts of Interest: The authors declare no conflicts of interest.

References

1. Staehelin, J.; Harris, N.R.; Appenzeller, C.; Eberhard, J. Ozone trends: A review. *Rev. Geophys.* **2001**, *39*, 231–290. [CrossRef]
2. Rabl, A.; Eyre, N. An estimate of regional and global O3 damage from precursor NOx and VOC emissions. *Environ. Int.* **1998**, *24*, 835–850. [CrossRef]
3. Lin, M.; Horowitz, L.W.; Xie, Y.; Paulot, F.; Malyshev, S.; Shevliakova, E.; Finco, A.; Gerosa, G.; Kubistin, D.; Pilegaard, K. Vegetation feedbacks during drought exacerbate ozone air pollution extremes in Europe. *Nat. Clim. Chang.* **2020**, *10*, 444–451. [CrossRef]
4. Monks, P.S.; Archibald, A.T.; Colette, A.; Cooper, O.; Coyle, M.; Derwent, R.; Fowler, D.; Granier, C.; Law, K.S.; Mills, G.E.; et al. Tropospheric ozone and its precursors from the urban to the global scale from air quality to short-lived climate forcer. *Atmos. Chem. Phys.* **2015**, *15*, 8889–8973. [CrossRef]
5. Martin, P.E.; Sanford, T. *Climate Change and Your Health: Rising Temperatures, Worsening Ozone Pollution*; Union of Concerned Scientists: Cambridge, MA, USA, 2011.
6. Gina, M.; Harmens, H. *Ozone Pollution: A Hidden Threat to Food Security*; NERC/Centre for Ecology & Hydrology: Bangor, UK, 2011.
7. Mills, G.; Sharps, K.; Simpson, D.; Pleijel, H.; Broberg, M.; Uddling, J.; Jaramillo, F.; Davies, W.J.; Dentener, F.; Berg, M.V.D.; et al. Ozone pollution will compromise efforts to increase global wheat production. *Glob. Chang. Biol.* **2018**, *24*, 3560–3574. [CrossRef]
8. Singh, A.A.; Agrawal, S.B. Tropospheric ozone pollution in India: Effects on crop yield and product quality. *Environ. Sci. Pollut. Res.* **2016**, *24*, 4367–4382. [CrossRef] [PubMed]
9. González-Fernández, I.; Bass, D.; Muntifering, R.; Mills, G.; Barnes, J. Impacts of ozone pollution on productivity and forage quality of grass/clover swards. *Atmos. Environ.* **2008**, *42*, 8755–8769. [CrossRef]
10. Xu, H.; Yang, J.C.; Chen, S.B.; Jiang, G.M.; Li, Y.G. Review of plant responses to ozone pollution. *Chin. J. Plant Ecol.* **2007**, *31*, 1205.
11. Tatarchenko, G.; Beloshitskaya, N.; Sychenko, V.; Liashuk, V. The effect of ozone on wet atmospheric corrosion of aluminium of high-voltage lines. *MATEC Web Conf.* **2019**, *294*. [CrossRef]

12. McDermid, I.S.; Godin, S.M.; Lindqvist, L.O. Ground-based laser DIAL system for long-term measurements of stratospheric ozone. *Appl. Opt.* **1990**, *29*, 3603–3612. [CrossRef]
13. Bhattacharya, A.B.; Bhoumick, A. The Antarctic Ozone Hole, midlatitudinal global ozone reduction and a comparison of ozone research to fundamental discoveries in Physics. *Sci. Cult.* **2011**, *77*, 226–228.
14. Liyana, A.N.; Abdullah, A.M. Ozone pollution and historical trends of surface background ozone level: A review. *World Appl. Sci. J.* **2011**, *14*, 31–38.
15. Clifton, O.E. Constraints on Ozone Removal by Land and Implications for 21th Century Ozone Pollution. Ph.D. Thesis, Columbia University, New York, NY, USA, 2018.
16. Cao, Y.; Xie, C.; Wang, B.; Cheng, L.; Fang, Z.; Li, L.; Zhuang, P.; Yang, H.; Shao, J. Design and optimization of Mie scattering lidar detection system. In *Sixth Symposium on Novel Optoelectronic Detection Technology and Applications*; SPIE: Bellingham, WA, USA, 2020; Volume 11455.
17. Hoff, R.M. Vertical structure of arctic haze observed by lidar. *J. Appl. Meteorol.* **1988**, *27*, 125–139. [CrossRef]
18. Mao, J.-D.; Hua, D.-X.; He, T.-Y.; Wang, M. Lidar observations of atmospheric aerosol optical properties over Yinchuan area. *Spectrosc. Spectr. Anal.* **2010**, *30*, 2006–2010.
19. Hua, D.; Uchida, M.; Kobayashi, T. Ultraviolet Rayleigh–Mie lidar for daytime-temperature profiling of the troposphere. *Appl. Opt.* **2005**, *44*, 1315–1322. [CrossRef]
20. Zhao, Y.; Lea, T.K.; Schotland, R.M. Correction function for the lidar equation and some techniques for incoherent CO_2 lidar data reduction. *Appl. Opt.* **1988**, *27*, 2730–2740. [CrossRef]
21. Measures, R.M.; Pilon, G. A study of tunable laser techniques for remote mapping of specific gaseous constituents of the atmosphere. *Opto-Electronics* **1972**, *4*, 141–153. [CrossRef]
22. Ahmed, S.A. Molecular air pollution monitoring by dye laser measurement of differential absorption of atmospheric elastic backscatter. *Appl. Opt.* **1973**, *12*, 901–903. [CrossRef]
23. Byer, R.L.; Garbuny, M. Pollutant detection by absorption using Mie scattering and topographic targets as retroreflectors. *Appl. Opt.* **1973**, *12*, 1496–1505. [CrossRef]
24. Browell, E.V.; Ismail, S.; Shipley, S.T. Ultraviolet DIAL measurements of O_3 profiles in regions of spatially inhomogeneous aerosols. *Appl. Opt.* **1985**, *24*, 2827–2836. [CrossRef]
25. Steinbrecht, W.; Carswell, A.I. Correcting for interference of Mt. Pinatubo aerosols on DIAL measurements of stratospheric ozone. In Proceedings of the Sixteenth International Laser Radar Conference, Cambridge, MA, USA, 20–24 July 1992; 1NASA, code JTT: Washington, DC, USA, 1992; Volume 19922.
26. Papayannis, A.; Ancellet, G.; Pelon, J.; Mégie, G. Multiwavelength lidar for ozone measurements in the troposphere and the lower stratosphere. *Appl. Opt.* **1990**, *29*, 467–476. [CrossRef]
27. McGee, T.J.; Gross, M.; Ferrare, R.; Heaps, W.; Singh, U. Raman dial measurements of stratospheric ozone in the presence of volcanic aerosols. *Geophys. Res. Lett.* **1993**, *20*, 955–958. [CrossRef]
28. Wang, Z.E.; Zhou, J.; Hu, H.L.; Gong, Z. Evaluation of dual differential absorption lidar based on Raman-shifted Nd: YAG or KrF laser for tropospheric ozone measurements. *Appl. Phys. B* **1996**, *62*, 143–147. [CrossRef]
29. Wang, Z.; Nakane, H.; Hu, H.; Zhou, J. Three-wavelength dual differential absorption lidar method for stratospheric ozone measurements in the presence of volcanic aerosols. *Appl. Opt.* **1997**, *36*, 1245–1252. [CrossRef] [PubMed]
30. Fernald, F.G. Analysis of atmospheric lidar observations: Some comments. *Appl. Opt.* **1984**, *23*, 652–653. [CrossRef] [PubMed]
31. Russell, P.B.; Bergstrom, R.W.; Shinozuka, Y.; Clarke, A.D.; DeCarlo, P.F.; Jimenez, J.L.; Livingston, J.M.; Redemann, J.; Dubovik, O.; Strawa, A. Absorption Angstrom Exponent in AERONET and related data as an indicator of aerosol composition. *Atmos. Meas. Tech.* **2010**, *10*, 1155–1169. [CrossRef]
32. Su, W.; Liu, C.; Hu, Q.; Fan, G.; Xie, Z.; Huang, X.; Zhang, T.; Chen, Z.; Dong, Y.; Ji, X.; et al. Characterization of ozone in the lower troposphere during the 2016 G20 conference in Hangzhou. *Sci. Rep.* **2017**, *7*, 17368. [CrossRef] [PubMed]
33. Yang, Z.-H.; Zhang, Y.-K.; Chen, Y.; Li, X.-F.; Jiang, Y.; Feng, Z.-Z.; Deng, B.; Chen, C.-L.; Zhou, D.-F. Simultaneous detection of multiple gaseous pollutants using multi-wavelength differential absorption LIDAR. *Opt. Commun.* **2022**, *518*, 128359. [CrossRef]
34. Ma, J.; Bi, J.; Li, B.; Zhu, D.; Wang, X.; Meng, Z.; Shi, J. Aerosol Vertical Structure and Optical Properties during Two Dust and Haze Episodes in a Typical Valley Basin City, Lanzhou of Northwest China. *Remote Sens.* **2024**, *16*, 929. [CrossRef]
35. Lei, L.; Berkoff, T.A.; Gronoff, G.; Su, J.; Nehrir, A.R.; Wu, Y.; Moshary, F.; Kuang, S. Retrieval of UVB aerosol extinction profiles from the ground-based Langley Mobile Ozone Lidar (LMOL) system. *Atmos. Meas. Tech.* **2022**, *15*, 2465–2478. [CrossRef]
36. Kuang, S.; Wang, B.; Newchurch, M.J.; Knupp, K.; Tucker, P.; Eloranta, E.W.; Garcia, J.P.; Razenkov, I.; Sullivan, J.T.; Berkoff, T.A.; et al. Evaluation of UV aerosol retrievals from an ozone lidar. *Atmos. Meas. Tech.* **2020**, *13*, 5277–5292. [CrossRef]

photonics

MDPI

Article

A Preliminary Study on the Principle of Linear Effect Scaling Laws for Laser Atmospheric Transmission

Xin Ye [1,2], Chengyu Fan [1,*], Wenyue Zhu [1], Pengfei Zhang [3], Xianmei Qian [1], Jinghui Zhang [1] and Tao Jiang [1]

1 Anhui Institute of Optics and Fine Mechanics, Hefei Institutes of Physical Science, Chinese Academy of Sciences, Hefei 230031, China; yx034020@163.com (X.Y.); zhuwenyue@aiofm.ac.cn (W.Z.); xmqian@aiofm.ac.cn (X.Q.); jhzhang@aiofm.ac.cn (J.Z.); tjiang@aiofm.ac.cn (T.J.)
2 University of Science and Technology of China, Hefei 230026, China
3 Institute of Intelligent Machines, Hefei Institutes of Physical Science, Chinese Academy of Sciences, Hefei 230031, China; pfzhang@aiofm.ac.cn
* Correspondence: cyfan@aiofm.ac.cn

Abstract: Numerical simulations were performed to rapidly predict and evaluate laser beam expansion caused by linear atmospheric transmission effects, such as turbulence and jitter, thereby enhancing the accuracy of the scaling law. Simulation results indicate that the turbulence term coefficient in the beam expansion calibration expression correlates linearly with the initial beam mass and inversely with the transmission distance. By fitting a nonlinear surface, the relationship between the turbulence term coefficient, initial beam mass, and transmission distance was established. Additionally, under turbulence-free conditions, a calibration expression relating initial beam mass to transmission distance was derived. The tracking jitter-term coefficient was determined to be 3.69, effectively characterizing beam expansion due to system jitter error. Based on simulation outcomes, a scaling law model for beam expansion induced by linear atmospheric transmission effects was clearly established. The model closely matched the simulation data, with a root mean square error (RMSE) of 3.88. Compared with existing scaling law simulations, the proposed calibration expression significantly enhances the accuracy in predicting and evaluating beam expansion caused by linear atmospheric transmission effects. It also provides a more precise characterization of variations in beam expansion during laser transmission.

Keywords: laser transmission; beam quality; turbulence effect; tracking jitter error; scaling law

1. Introduction

Lasers, as critical light sources, have attracted significant attention due to their atmospheric transmission characteristics. When laser beams propagate through the atmosphere, they experience linear effects including absorption, scattering, and turbulence, as well as nonlinear thermal halo effects, all leading to beam spot expansion [1–3]. This expansion degrades the beam quality, significantly affecting performance in laser-based atmospheric applications such as communications, remote sensing, and range measurements. Understanding and modeling laser atmospheric transmission requires fundamental physical principles, mathematical models, complex numerical simulations, and experimental verification. By establishing accurate calibration relationships, researchers can quickly predict and evaluate laser beam expansion under varying environmental conditions.

In recent years, with the development of computer technology, based on the fluctuating optics theory, the numerical simulation method has become an important research tool for studying the variation in the quality of laser beams transmitted in the atmosphere. Although fluctuating optics simulations can accurately reproduce the spatial and temporal variations in atmospheric beam transmission, they require consideration of multiple factors, such as initial beam quality, atmospheric conditions, system jitter, and blocking ratio [4,5]. Consequently, their slow computational speed limits their practicality for real-world applications [6]. Based on the analysis method of mean square and radius [7], domestic scholars have successively proposed a variety of calibration laws, including the atmospheric transmission calibration law of focused Gaussian beam [8] and focused platform beam [9–13], the calibration law of genetic algorithm optimization [6], and the improved numerical model of the calibration law [6,14], etc., which mainly use the control variables to analyze, one by one, the change in expression coefficients induced by the change in individual parameters. On the basis of determining the coefficients of one expression, the value of the coefficients of another expression is further determined by changing other parameters, and the calibration expression is finally determined. A limitation of this method is its inability to traverse multiple parameter combinations, thereby failing to accurately characterize laser transmission under diverse atmospheric conditions. If expression coefficients are merely averaged from specific scenarios, broader applicability and enhanced accuracy of the calibration expressions become challenging to achieve [14,15]. Additionally, comparisons of previous research results indicate significant variations in the calibration expression coefficients depending on parameters such as beam type, initial beam quality, transmission distance, and system jitter error [10]. Thus, employing constant values to characterize these coefficients does not accurately reflect actual conditions.

This study utilizes the mean square sum relationship to analyze linear effects in atmospheric transmission. Numerical simulations were conducted by systematically varying parameter combinations to investigate laser transmission characteristics under diverse atmospheric conditions. Relationships between the calibration expression coefficients and variables including initial beam quality, transmission distance, and system jitter error were established. This approach significantly enhances the accuracy of calibration expressions describing linear effects in laser atmospheric transmission.

2. Theoretical Analysis

Typically, changes in laser beam quality in the far field after atmospheric transmission can be assessed using parameters such as the beam quality factor, which quantifies beam expansion at the focal plane. It is commonly assumed that the various perturbation sources contributing to beam expansion act independently [12]. When considering linear effects such as laser system jitter, atmospheric absorption, scattering, and turbulence, it is typically assumed these effects combine at the beam's focal plane. Specifically, the squared spot radius equals the sum of the squares of the effective radii from each individual effect, satisfying the mean square sum relationship:

$$\beta^2 = \beta_0^2 + \beta_T^2 + \beta_J^2, \quad (1)$$

where β is the beam quality factor after beam expansion due to linear effects of laser transmission, β_0 is the beam quality factor characterizing expansion due to diffraction, β_T is the beam quality factor characterizing expansion due to turbulence, and β_J is the beam quality factor characterizing expansion due to tracking jitter.

According to the literature [1,11], Equation (1) can be written as follows:

$$\beta^2 = \beta_0^2 + A(D/r_0)^2 + B(\sigma_i/\sigma_0)^2, \quad (2)$$

where D is the laser emission aperture; σ_i is the system tracking jitter error; σ_0 is the diffraction angle of the emission system; and r_0 is the atmospheric coherence length. The first and second terms on the right side of the equal sign in Equation (2) characterize beam expansion due to diffraction and turbulence [4], while the third term characterizes beam expansion due to tracking jitter.

The diffraction angle σ_0 of the emitting system is given by the following:

$$\sigma_0 = 1.22\lambda/D, \tag{3}$$

The expression for the atmospheric correlation length r_0 is given by the following:

$$r_0 = \left[0.423k^2 \int_0^L C_n^2(z)(1 - z/L)^{5/3}\mathrm{d}z\right]^{-3/5}, \tag{4}$$

where λ is the emitted laser wavelength, k is the wave number, and C_n^2 is the refractive index structure constant.

Since it is not very reasonable to characterize the beam expansion rate with a constant, a relational equation for the beam expansion law of laser atmospheric transmission containing linear effects such as diffraction, turbulence, and tracking jitter is proposed:

$$\beta^2 = \beta_0^2 + A_m(D/r_0)^2 + B_m(\sigma_i/\sigma_0)^2, \tag{5}$$

where A_m and B_m take values as a function of parameters related to the laser transmission distance L and initial beam quality β_0, among others. For the proposed calibration relation (5), a numerical simulation is performed using a four-dimensional program to verify its feasibility and accuracy.

3. Numerical Simulation Results and Analysis

In the numerical simulation calculation, the wavelength is a truncated Gaussian beam of 1 μm, the laser emission aperture is 0.5 m to 1m, the initial beam quality is 1 to 10, the horizontal atmospheric transmission distance is 1 km to 9 km, the tracking jitter error of the laser system is 2.5 μrad, 5 μrad, and 7.5 μrad, and the refractive index structural constant takes values in the range of $1 \times 10^{-16}\ \mathrm{m}^{-2/3}$ to $1 \times 10^{-14}\ \mathrm{m}^{-2/3}$. The average wind speed is 2 m/s, the laser transmission time is 10 s, phase screen generation is performed via the spectral inversion method [16,17], the number of phase screen calculation grids is 256×256, the number of transmission steps, i.e., the number of phase screens, is 50, and the statistical results are taken from 30 laser transmission long exposures.

Without considering system jitter conditions, the beam expansion satisfies the calibration relation $\beta^2 = \beta_0^2 + A_m(D/r_0)^2$, and the variation in the expansion multiple $\beta^2 - \beta_0^2$ with the turbulence term $(D/r_0)^2$ for 63.2% of the ring-envelope energy radius in the focal plane caused by the turbulence effect is given in Figure 1. The variation in the turbulence term $(D/r_0)^2$ in the figure is calculated by changing the parameters of the initial beam quality β_0, transmission distance L, and turbulence intensity (C_n^2), where D/r_0 takes the value in the range of 0.866–19.295 and is fitted with a k value of about 1 using the linear expression $y = k \cdot x$, so that $A_\mathrm{m} \approx 1$. In this paper, the main simulation input parameters are transmission distance, initial beam quality, and atmospheric turbulence intensity, etc. These will be traversed in combinations; i.e., all laser transmission situations under the setup conditions are considered, totaling 1800 combinations. Because the initial beam quality β_0 changes will not cause the atmospheric correlation length to change, and under the same atmospheric turbulence intensity conditions, the transmission distance is determined, the size of the $(D/r_0)^2$ term remains unchanged, while β^2 increases with the increase of

β_0^2. Consequently, Figure 1 appears with $(D/r_0)^2$ taking a certain value, where $\beta^2 - \beta_0^2$ corresponds to a number of groups of values. The value of this phenomenon is mainly due to the value of β_0 caused by the different value of the phenomenon.

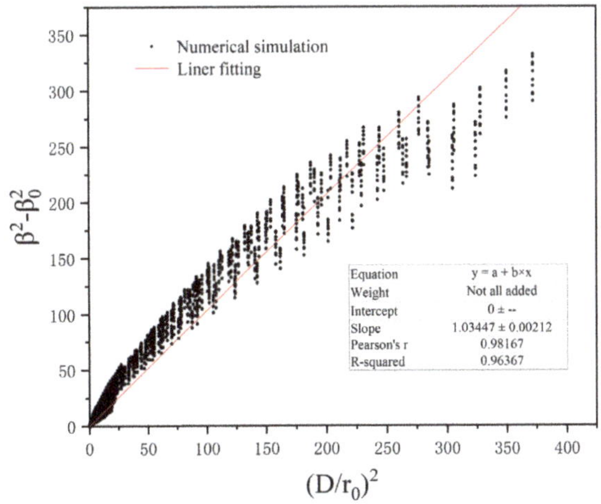

Figure 1. The variation in coefficient A with the initial beam quality.

The Amplitude Scintillation parameter, $A_m \approx 1$, can characterize the trend of spot expansion caused by the turbulence effect, but there is a large error, which is mainly manifested in the large difference in the A_m values obtained from the fits with different initial beam qualities β_0 or different transmission distances L. To observe the difference in the A_m values obtained from the fits more intuitively, we take β_0 = 1 and 8, and L = 1 km and 9 km, respectively. By changing the turbulence strength, the variation relationship of the expansion multiplier $\beta^2 - \beta_0^2$, with the turbulence term $(D/r_0)^2$, is obtained as shown in Figure 2a,b, and the fitting results are shown in Table 1.

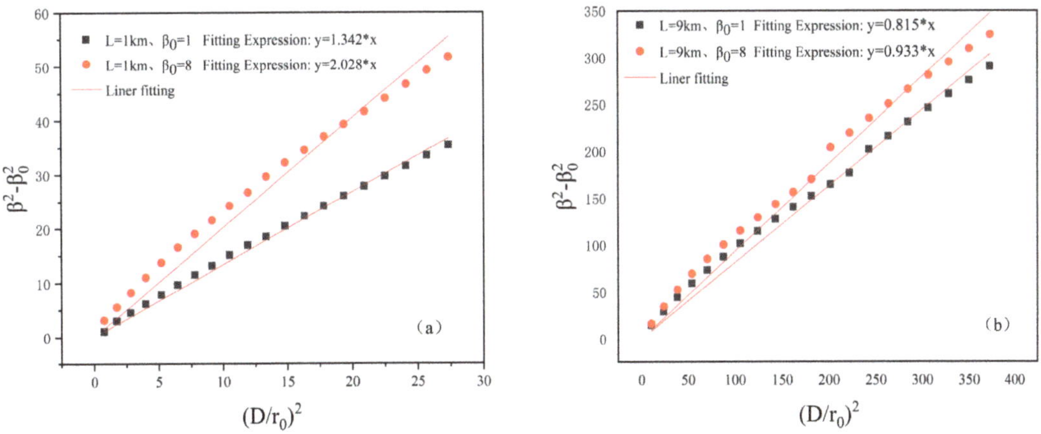

Figure 2. The initial beam quality is 1 and 8, and beam broadening caused by turbulence effects is observed when the laser propagates through the atmosphere over distances of 1 km and 9 km: (**a**) Under different initial beam quality conditions, the variation in beam expansion caused by turbulence effects after laser transmission over 1 km. (**b**) Under different initial beam quality conditions, the variation in beam expansion caused by turbulence effects after laser transmission over 9 km.

Table 1. The fitting values of coefficient A_m for β_0 and L under different value conditions.

A_m	β_0=1	β_0=8
L = 1 km	1.34	2.03
L = 9 km	0.81	0.93

When the initial beam quality β_0 is 1 and the transmission distance L is 1 km, the fitting coefficient A_m is 1.34. As β_0 increases to 8, Am decreases to 2.03, indicating that A_m increases with β_0 under the same distance conditions. When β_0 is 8 and L is 9 km, A_m is 0.93. As L decreases to 1 km, A_m increases to 2.03, indicating that A_m increases with decreasing L under the same initial beam quality conditions. Additionally, for initial beam qualities of 1 and 8 and laser atmospheric transmissions of 1 and 9 km, the A_m coefficient values range from about 0.814 to 2.03, with a difference of about 2.5 times between the maximum and minimum values. Using $A_m \approx 1$ to characterize the values of A_m in this range would result in a large error.

The coefficient A_m characterizes the rate of change in the turbulence term $(D/r_0)^2$ from expression $\beta^2 = \beta_0^2 + A_m(D/r_0)^2$. Comparing Figure 2a,b, A_m decreases with increasing β_0 and increases with decreasing transmission distance L. Therefore, the magnitude of A_m is related to the values of β_0 and L. To better understand the relationship between A_m, β_0, and L, under the same wavelength and aperture conditions, we traverse the initial beam quality β_0 and transmission distance L, taking a value combination. Adjusting the intensity of atmospheric turbulence, with these parameters as input, we simulate the coefficient A_m and obtain the corresponding relationships of β_0 and L. In Figure 3, we show the coefficient A_m and its corresponding relationship with β_0, and in Figure 4, we show the coefficient A_m and its corresponding relationship with L.

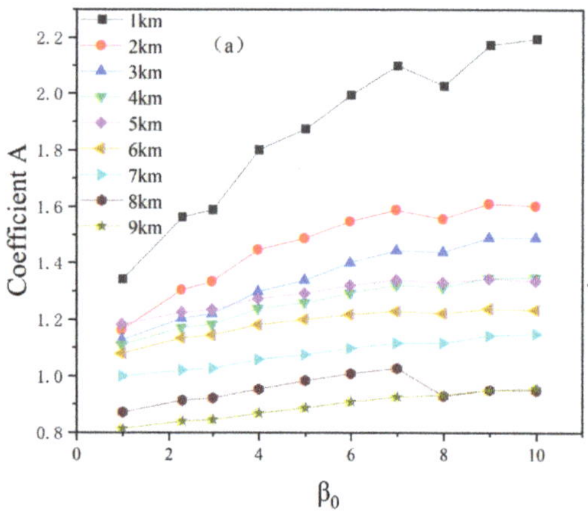

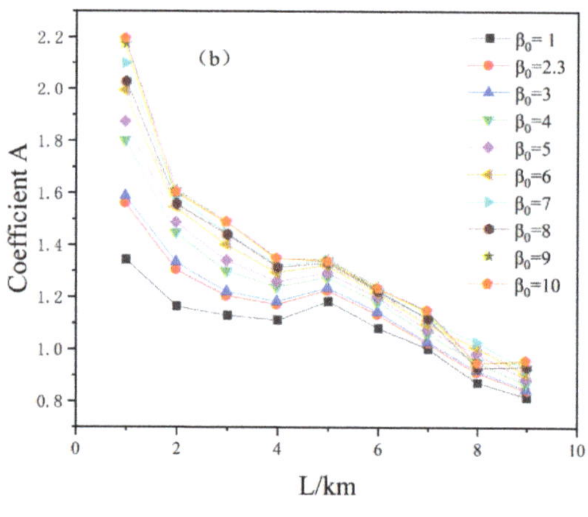

Figure 3. Relationship between the coefficient A_m and β_0, L. (**a**) Variation in the coefficient A_m with β_0; (**b**) variation in the coefficient A_m with L.

Beam quality is typically characterized by parameters such as beam divergence, divergence angle, M^2 factor (aberration), and beam quality factor. All factors affecting near-field beam quality also impact far-field beam quality [18]. A laser beam with high beam quality ensures a small beam waist and long Rayleigh length, enabling transmission over longer distances [19]. As expressed by $\beta^2 = \beta_0^2 + A_m(D/r_0)^2$, the coefficient A_m mainly characterizes the rate of change in the turbulence term $(D/r_0)^2$. When the initial beam quality is small and the beam divergence is low with a small cross-sectional area, the beam expansion caused by the turbulence effect is reduced to a certain extent, protecting the energy density in the beam center and maintaining it constant during propagation or slowly decreasing. The turbulence effect induced by beam expansion changes slowly, with the coefficient A_m being smaller. When the initial beam quality is larger, the beam divergence is higher, and the cross-sectional area is larger; the beam is more susceptible to turbulence during transmission, resulting in increased beam expansion, deteriorated beam quality, and rapidly changing beam expansion induced by the turbulence effect, with the coefficient A_m increasing. As shown in Figure 3, with the increase of β_0, the turbulence-induced beam expansion is intensified, corresponding to the increasing coefficient A_m and the overall linearly increasing relationship, satisfying $A_m \propto \beta_0$, which can be expressed as

$$A_m = a + b \cdot \beta_0 \tag{6}$$

which characterizes the changing relationship between A_m and β_0, and a and b are the fitting parameters.

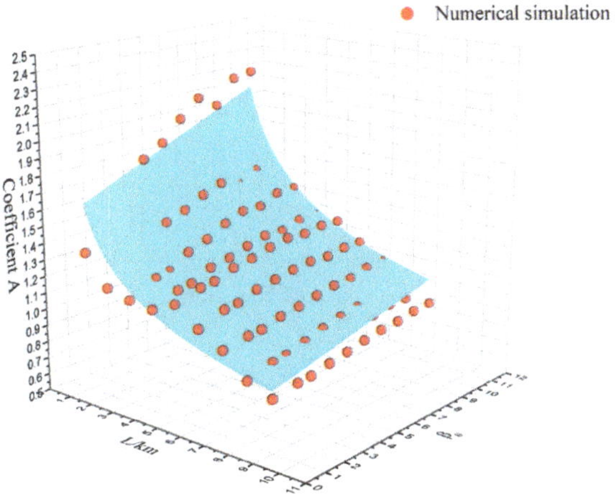

Figure 4. Schematic diagram of nonlinear surface fitting.

Furthermore, as the transmission distance increases, the laser beam experiences more turbulence effects, and the beam expansion caused by the turbulence effect accumulates; at this time, it is difficult to cause a rapid change in the total cumulative effect of the turbulence effect by changing the characteristic parameter D/r_0 of the turbulence effect; i.e., after long-distance transmission, the rate of change in the beam expansion caused by the turbulence effect decreases gradually. Combined with Figure 4, it can be seen that under different initial beam qualities β_0, the coefficient A_{m} decreases with the increase in the transmission distance L, and the overall relationship is an inverse proportional decreasing relationship, satisfying $A_m \propto 1/L^n$ $(n > 0)$, which can be expressed as

$$A_m = c/L^n (n > 0) \tag{7}$$

which characterizes the changing relationship between Am and L, and c is the fitting parameter.

On this basis, a new expression can be constructed:

$$A_m = \frac{a + b \cdot \beta_0}{L^n}, n > 0 \tag{8}$$

It describes the functional relationship between the coefficient A_m and the initial beam quality β_0, and the transmission distance L. By nonlinear surface fitting, the fitting results are obtained as shown in Figure 4. The fitting coefficient of determination, $R^2 = 0.91$, indicates that there is a high degree of agreement between the distribution of data points and the fitted surface, providing a good explanation of the model.

Based on the fitting results depicted in Figure 4, the fitted expression is obtained as follows:

$$A_m = \frac{1.58 + 0.052 \cdot \beta_0}{L^{0.291}} \tag{9}$$

The fitting parameters are as follows: a = 1.58, b = 0.052, $n = 0.291$, and the length scale L in kilometers.

Considering the system tracking jitter error, the turbulence intensity is set to 0; i.e., the influence of turbulence is not considered, and the numerical simulation results use the statistical values of 30 laser transmission long exposures. From Equation (3), it can be seen that the system tracking jitter error caused by the beam expansion term calculation is mainly related to the wavelength and launching aperture, and is not related to the transmission distance. Therefore, σ_0 can be calculated to be 1.622 μrad under the conditions of a wavelength of 1 μm and a launching aperture of 0.7 m. The simulation calculates the initial beam quality β_0 to be 1, 3, 5, 7, and 9, and the tracking jitter error σ_j to be 2.5 μrad, 5 μrad, and 7.5 μrad under far-field beam quality conditions. Fitting these results yields different values for coefficient B, as shown in Table 2. The average value of B is 3.69.

Table 2. The fitting values of coefficient B_m for β_0 and σ_j under different value conditions.

σ_i/μrad		**2.5**	**5**	**7.5**	
B_m	$\beta_0 = 1$	3.265	3.270	3.285	$\overline{B}_m = 3.69$
	$\beta_0 = 3$	3.980	3.702	3.491	
	$\beta_0 = 5$	3.532	3.737	3.684	
	$\beta_0 = 7$	3.639	3.825	3.773	
	$\beta_0 = 9$	4.338	3.996	3.877	

As demonstrated in Table 2, under identical initial beam quality conditions, the rate of beam extension variation induced by system tracking jitter error remains nearly constant. This observation suggests a steady growth pattern of beam extension caused by the jitter. Notably, although an increase in initial beam quality leads to a gradual rise in the beam expansion rate, the fluctuation range of this rate remains centered around 3.69, indicating minimal sensitivity to initial beam quality. Furthermore, under turbulence-free conditions with a fixed system tracking jitter error (e.g., 2.5 μrad), the relative contribution of beam expansion attributed to this error decreases significantly—from 88.6% to 11.3%—as the initial beam quality increases from 1 to 9. This inverse relationship implies that higher initial beam quality diminishes the proportional impact of jitter-induced beam extension on the total expansion. When turbulence effects are considered, the influence of the error coefficient B becomes even less pronounced. Consequently, the parameter $\overline{B}_m = 3.69$ effectively characterizes the beam expansion dynamics governed by system jitter error.

In conclusion, the linear atmospheric propagation effects on beam spreading at the $1/e^2$ intensity contour demonstrate quantifiable scaling relationships for $\lambda = 1$ μm laser systems. Through systematic parametric analysis (D = 0.7 m transmitter aperture, L = 1–9 km propagation distance, $\beta_0 = 1$–10 initial beam quality factor), experimental measurements confirm that the radial spread satisfies the dimensionless calibration model:

$$\beta^2 = \frac{1.58 + 0.052\beta_0}{L^{0.291}}(D/r_0)^2 + 3.69(\sigma_j/\sigma_0)^2 \tag{10}$$

This finding aligns with the turbulence-independent scaling law proposed by Andrews et al. [20] while extending its applicability to low-coherence beams ($\beta_0 > 5$) through modified terms.

To validate the predictive accuracy of Equation (10), we conducted numerical simulations using a Monte Carlo approach (N = 500 iterations) with the following constrained parameters: propagation distance $L = 2$ km, initial beam quality factor $\beta_0 = 5$, and pointing jitter $\sigma_j = 2.5$ μrad. The simulated beam spreading characteristics were systematically compared with both the proposed calibration model and established scaling laws from Refs. [1,14], please refer to Appendix A. Under strong turbulence conditions, the laser beam undergoes breakup into multiple sub-spots, whose statistical properties are governed by the turbulence inner scale (l_0) and non-Kolmogorov spectral characteristics. If discrepancies exist in the phase screen generation algorithms or the truncation of scattering orders (e.g., neglecting higher-order scattering terms), the turbulence-induced beam breakup effect will be underestimated, thereby introducing deviations in the simulated beam spreading and intensity statistics.

As evidenced in Figure 5, both Equation (10) and the calibration models from Refs. [1,14] demonstrate satisfactory agreement with numerical simulations. Specifically, Equation (10) achieves the minimal deviation from simulated results with a root mean square error (RMSE) of 3.88. With increasing turbulence intensity ($C_n^2 > 10^{-14}$ m$^{-2/3}$), the prediction accuracy of Refs. [1,14] deteriorates significantly, yielding RMSE values of 10.91 and 14.87, respectively. This comparative analysis confirms that the proposed calibration formalism in Equation (10) enhances prediction accuracy for laser beam spreading by 64.4% and 73.8% relative to Refs. [1,14], achieved through systematic parameter space exploration ($L \in [0.5, 10]$ km, $\beta_0 \in [1, 10]$, $C_n^2 \in [10^{-16}, 10^{-13}]$ m$^{-2/3}$). The improved fidelity originates from optimized weight coefficients in the generalized scaling law, which effectively minimizes overfitting through global sensitivity analysis across 1800 parameter combinations.

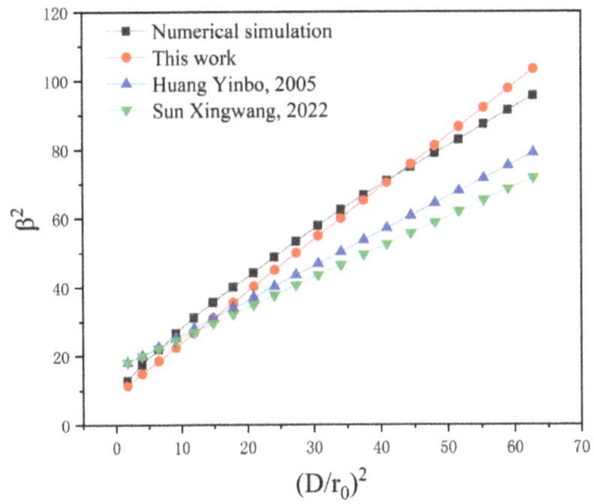

Figure 5. Comparison of fitting effects of different calibration expressions [1,14].

4. Conclusions

This study systematically investigates laser beam spreading dynamics caused by linear atmospheric effects, including diffraction, turbulence, and jitter, through detailed numerical simulations. Quantitative analysis demonstrates how beam expansion varies parametrically with initial beam quality (M^2 = 1–10) and propagation distance (L = 1–9 km), given a fixed transmitter aperture (D = 0.5–1 m) at a wavelength of $\lambda = 1$ μm. A generalized calibration model was established through global optimization across 1800 parameter combinations, resulting in excellent agreement with simulated beam profiles within the $1/e^2$ intensity contour. Comparative validations show a 64.2% reduction in root mean square deviation (RMSE = 3.88) compared to established models [1,14]. This improvement is especially significant under strong turbulence conditions ($Cn^2 > 10^{-14}$ $m^{-2}/^3$), where previous formulations exhibit a 12–22% overestimation.

Although this study focuses on horizontal propagation paths to isolate turbulence effects, two critical extensions are proposed: (1) implementation of slant-path transmission models incorporating altitude-dependent profiles consistent with the Hufnagel–Valley (HV) atmospheric model; and (2) integration of nonlinear thermal blooming effects via a coupled solution of wave propagation and heat diffusion equations.

Author Contributions: Supervision and conceptualization, C.F. and W.Z.; resources, C.F.; software, J.Z. and T.J.; methodology, P.Z. and X.Q.; numerical simulation and writing—original draft preparation, X.Y.; writing—review and editing, X.Y. All authors have read and agreed to the published version of the manuscript.

Funding: This research received no external funding.

Data Availability Statement: The data mentioned in the manuscript may be requested by email from the corresponding author.

Acknowledgments: We would like to thank the Anhui Institute of Optics and Fine Mechanics at Hefei Institute of Physical Science for providing numerical simulation guidance.

Conflicts of Interest: The authors declare no conflict of interest.

Appendix A

The influence of tracking jitter in laser transmission systems on beam broadening during atmospheric propagation through turbulence has been numerically investigated in Ref. [1]. Their results demonstrated the following scaling relationship for the 63.2% encircled energy radius:

$$\beta^2 = \beta_0^2 + (D/r_0)^2 + 6.8(\sigma_i/\sigma_0)^2$$

Meanwhile, Ref. [14] proposed an integrated model describing the 63.2% encircled power radius at the focal plane under combined effects of diffraction, turbulence, tracking jitter, and thermal blooming, expressed as follows:

$$\beta^2 = \beta_0^2 + \left\{0.0043\exp(\lambda) + \left[(\beta_0/10.2)^{6.1} + 0.86^2\right]^{1/2}\right\}(D/r_0)^2 + 6.93(\sigma_i/\sigma_d)^2$$

Both studies primarily employed a controlled variable approach to sequentially analyze coefficient variations induced by individual parameter adjustments. Specifically, they determined one coefficient of the scaling expression while fixing other parameters and then iteratively resolved subsequent coefficients through similar procedures. However, this methodology inherently limits the exploration of multidimensional parameter space combinations.

To address this limitation, our work implements comprehensive numerical simulations encompassing systematically varied combinations of laser transmission parameters. Through this parametric sweep strategy, we establish functional relationships between the scaling coefficients and key parameters including initial beam quality, propagation distance, and system jitter error. This systematic approach significantly enhances the accuracy of scaling law formulations for linear effects in laser atmospheric propagation by explicitly considering multidimensional parameter interdependencies.

References

1. Huang, Y.B.; Wang, Y.J. The Effect of Tracking Jitter on the Beam Spreading Induced by Atmospheric Turbulence. *Chin. J. Acta Opt. Sin.* **2005**, *25*, 152–156.
2. Su, Y.; Wan, M. (Eds.) *High-Energy Laser Systems (Gaoneng Jiguang Xitong)*; National Defense Industry Press: Beijing, China, 2004.
3. Rao, R.Z. *Modern Atmospheric Optics (Xiandai Daqi Guangxue)*; Science Press: Beijing, China, 2012.
4. Shi, X.Y.; Wang, Y.J.; Huang, Y.B. Numerical analysis on the uniform focused beam spreading induced by atmosphere turbulence with different obscuring ratio. *Chin. J. High Power Laser Part. Beams* **2003**, *15*, 1181–1183.
5. Shi, X.Y.; Wang, Y.J. Experimental and Numerical Analysis on the Focused Beam Spreading at Different Obscuring Ratio. *Chin. J. Hangzhou Dianzi Univ.* **2005**, *25*, 94–96.
6. Chen, X.; Zhu, W.; Qian, X.; Wu, P.; Qing, C.; Sun, G.; Wei, H.; Weng, N.; Cui, X. Scale Model of Focused Gaussian Beam Propagating in Turbulent Atmosphere. *Chin. J. Lasers* **2023**, *50*, 109–118.
7. Breaux, H.; Evers, W.; Sepucha, R.; Whitney, C. Algebraic Model for CW Thermal-Blooming Effects. *Chin. J. Appl. Phys.* **1975**, *46*, 1–7. [CrossRef] [PubMed]
8. Stock, R.D. High energy laser scaling laws. In Proceedings of the 2003 Directed Energy Modeling and Simulation Conference, Albuquerque, NM, USA, 25–27 March 2003.
9. Qiao, C.H.; Fan, C.Y.; Wang, Y.J. Scaling laws about DF high energy laser propagation in real atmosphere. *Chin. J. High Power Laser Part. Beams* **2007**, *19*, 1965–1969.
10. Hu, H.Y.; Yang, B.W.; Zhong, Z.Q.; Zhang, B. Beam Spreading Scaling Law of Gaussian Beam Propagating in Anisotropic Non-Kolmogorov Maritime Atmospheric Turbulence. *Chin. J. Acta Photonica Sinica* **2025**, *54*, 0130004.
11. Wang, Y.J.; Huang, Y.B. Analysis of the Scaling Laws about Focused Uniform Beam Spreading Induced by Real Atmosphere. *Chin. J. Quantum Electron.* **2006**, *23*, 274–281.
12. Qiao, C.H.; Fan, C.Y.; Huang, Y.B.; Wang, Y. Scaling Laws of High Energy Laser Propagation through Atmosphere. *Chin. J. Lasers* **2010**, *37*, 433–437. [CrossRef]
13. Huang, Y.B.; Wang, Y.J. Numerical analysis of the scaling laws about focused beam spreading induced by the atmosphere. *Chin. J. Acta Phys. Sin.* **2006**, *55*, 6715–6719. [CrossRef]
14. Sun, X.W.; Zhang, Q.; Zhong, Z.Q.; Zhang, B. Scaling Law for Beam Spreading During High-Energy Laser Propagation in Atmosphere. *Chin. J. Acta Opt. Sin.* **2022**, *42*, 74–80.
15. Zhang, Q.; Sun, X.W.; Zhong, Z.Q.; Zhang, B. Scaling Laws of Beam Spreading in Laser Ocean Turbulence. *Chin. J. Acta Opt. Sin.* **2024**, *44*, 92–99.
16. Lane, R.G.; Glindemann, A.; Dainty, J.C. Simulation of a kolmogorov phase screen. *Chin. J. Waves Random Media* **1992**, *2*, 209–224. [CrossRef]
17. Zhang, H.; Li, X. Numerical simulation of wavefront phase screen distorted by atmospheric turbulence. *Chin. J. Opto-Electron. Eng.* **2006**, *33*, 14–19.
18. Du, X.W. Factors for Evaluating Beam Guality of a Real High Power Laser on the Target Surface in Far Field. *Chin. J. Lasers* **1997**, *24*, 40–45.
19. Chen, H. *Influence of Beam Quality on Propagation and Focus of Laser Beam and Quality of Laser Materials Processing*; Beijing University of Technology: Beijing, China, 2006.
20. Andrews, L.C.; Phillips, R.L. *Laser Beam Propagation Through Random Media*, 2nd ed.; SPIE Press: Bellingham, WA, USA, 2005; pp. 1–783. [CrossRef]

photonics

MDPI

Article

Reconstruction Algorithm of Absorption Spectral Field Distribution Based on a Priori Constrained Bivariate Polynomial Model

Chuge Chen [1,2], Dingfeng Shi [1,2], An Huang [1], Suman Ai [1,2], Rantong Niu [1,2], Ting Jiao [1] and Zhenyu Xu [2,*]

[1] Hefei Institutes of Physical Science, Chinese Academy of Sciences, Hefei 230031, China; cgchen@aiofm.ac.cn (C.C.)
[2] Science Island Branch, Graduate School of USTC, Hefei 230026, China
[*] Correspondence: zyxu@aiofm.ac.cn

Abstract: Computed Tomography–Tunable Diode Laser Absorption Spectroscopy (CT-TDLAS) is an effective diagnostic method for analyzing combustion flow fields within engines. This study proposes an adaptive reconstruction algorithm utilizing constrained polynomial fitting within the CT-TDLAS framework. Based on existing polynomial fitting models, the proposed algorithm integrates physical boundary constraints on temperature and concentration fields, optimizing integrated absorbance errors. This method significantly enhances reconstruction accuracy and computational efficiency, while also lowering computational complexity. The adaptive strategy dynamically adjusts the polynomial order, effectively mitigating issues of overfitting or underdetermination typically associated with fixed polynomial orders. Numerical simulations demonstrate reduced temperature reconstruction errors of 2%, 1.6%, and 2% for single-peak, dual-peak, and mixed distribution flow fields, respectively. Corresponding concentration errors were 2%, 1.8%, and 2.6%, which are all improvements over those achieved by the Algebraic Reconstruction Technique (ART). Experimental results using a McKenna flat-flame burner revealed an average reconstruction error of only 0.3% compared to thermocouple measurements for high-temperature regions (>1000 K), with a minimal central temperature difference of 6 K. For lower-temperature peripheral regions, the average error was 188 K, illustrating the practical applicability of the proposed algorithm.

Keywords: tunable diode laser absorption spectroscopy (TDLAS); combustion flow field diagnosis; constrained polynomial fitting; adaptive strategy

1. Introduction

With the ongoing advancement in aerospace technology, accurately diagnosing combustion flow fields in engines, including aircraft and scramjets, has become critically important. Traditional diagnostic tools, such as pressure sensors, intrusive probes, and thermocouples, typically disrupt flow, exhibit slow responses and low sensitivity, and are unsuitable for harsh, high-temperature, and high-speed environments typical of hypersonic vehicles. TDLAS technology, employing molecules such as H_2O, CO_2, CO, and O_2 as probes [1–4], enables non-intrusive, rapid, and real-time measurements of temperature, pressure, flow velocity, and gas concentration in engine combustion flow fields. This technique robustly supports aerodynamic studies of burners, key component experiments, and ground-based performance evaluations [5–8]. However, since TDLAS measurements provide line-of-sight averages, obtaining two-dimensional flow field parameter distributions necessitates multi-beam and multi-angle measurement approaches. By integrating absorption spectral data obtained from multiple laser paths with Computed Tomography (CT) algorithms, two-dimensional reconstructions of flow field parameters can be effectively achieved, as exemplified by the CT-TDLAS method.

Conventional CT-TDLAS algorithms typically use linear tomographic methods, such as the Algebraic Reconstruction Technique (ART), to reconstruct integrated absorbance data across multiple wavelengths. Subsequently, they employ two-line ratio or Boltzmann plot methods to determine temperature and concentration distributions. For instance, Bao et al. [9] validated an enhanced two-line ratio method using numerical simulations and conceptual experiments. Xu et al. [10] applied various Radial Basis Function (RBF) models to fit integrated absorbance data across multiple wavelengths. They subsequently utilized a two-line ratio method for temperature and concentration reconstruction, obtaining results that closely matched the original distributions. However, linear tomographic algorithms, which ignore correlations among different wavelengths, often suffer from inaccuracies due to insufficient projections [11] and noise interference [9]. This frequently results in edge distortions within complex flow fields. Nonlinear CT-TDLAS methods, such as simulated annealing (SA), jointly process multi-wavelength absorption information to find optimal solutions [12,13]. However, these methods typically require excessive computational time. Recent studies have investigated deep neural networks to improve reconstruction capabilities [14–16]. Nevertheless, the accuracy of these models heavily depends on the quality of training data, and their generalizability to different combustion configurations remains uncertain. Additionally, Wang et al. [17] employed polynomial models to fit temperature and concentration distributions, resolving these based on residuals from multi-wavelength absorbance curves. This method uses multi-objective optimization of absorbance calculations and achieves high reconstruction accuracy. Currently, reconstruction algorithms based on absorbance calculations exhibit high computational complexity and introduce lineshape errors during processing. Furthermore, determining the optimal polynomial order to balance descriptive accuracy and computational efficiency remains unaddressed. Thus, developing an efficient, robust, and scalable reconstruction algorithm remains a critical challenge in this field.

In this study, we introduce a CT-TDLAS reconstruction algorithm utilizing constrained polynomial fitting. The algorithm optimizes integrated absorbance errors by incorporating physical boundary constraints for temperature and concentration fields. Additionally, we develop an adaptive strategy for selecting the polynomial order. This strategy dynamically adjusts polynomial orders according to the complexity of the flow field, effectively mitigating problems such as overfitting and underdetermination. Numerical simulations demonstrate that the temperature reconstruction errors remain below 3% for various flow field configurations. This represents an improvement of more than 50% compared to the traditional ART algorithm. Experiments further confirm that the reconstruction results of the algorithm are consistent with the measurement trend of the thermocouple.

2. Materials and Methods

2.1. Principles of CT-TDLAS

CT-TDLAS combines tunable diode laser absorption spectroscopy (TDLAS) with computer tomography (CT) to achieve noncontact spectral measurements of two-dimensional fields. TDLAS employs a tunable laser to generate a beam over a specific wavelength range, which, upon passing through the test gas, undergoes absorption at characteristic wavelengths. The transmitted beam is then detected, and the obtained spectral information is related to the gas species, temperature, concentration, and pressure. According to the Beer–Lambert law, the relationship is given by

$$\alpha_v = -ln(\frac{I_t}{I_0}) = \int_0^L C(l)\cdot P(l)\cdot S[T(l)]\cdot \phi dl \tag{1}$$

where α_v is the absorbance; I_0 is the incident light intensity in the absence of absorption; I_t is the transmitted light intensity after absorption; $C(l)$ is the concentration of the target gas species; $P(l)$ [atm] represents the pressure in the absorption region; $S[T(l)]$ [cm^{-2} atm^{-1}] is the line strength dependent on temperature; ϕ [cm] is the lineshape function; and L [cm] is the absorption path length. Because the lineshape function is normalized, the integral absorption A_v can be described as

$$A_v = \int_0^L C(l)\cdot P(l)\cdot S[T(l)]\cdot \phi dl \tag{2}$$

Using this principle, TDLAS technology can retrieve spectral information for specific absorption lines, which in turn can be used to infer gas temperature and concentration [18,19]. By combining data from multiple angles and laser paths with CT reconstruction algorithms, a two-dimensional map of flow field parameters is obtained [20,21].

2.2. Constrained Polynomial Fitting Reconstruction Algorithm

The constrained polynomial fitting (CPF) algorithm for reconstructing combustion fields begins by discretizing the measurement domain into pixels, each carrying independent temperature and concentration information. Since the distributions of temperature and concentration vary continuously over the region, a bivariate polynomial in spatial coordinates can be used as the basis function to approximate these distributions [17]. Assuming a maximum polynomial order m, the temperature and concentration fields are expressed as

$$T(x,y) = \sum_{k=0}^{m}\sum_{l=0}^{k} b_{k-l,l}x^{k-l}y^l = F_T(b_{kl}) \tag{3}$$

$$C(x,y) = \sum_{k=0}^{m}\sum_{l=0}^{k} a_{k-l,l}x^{k-l}y^l = F_c(a_{kl}) \tag{4}$$

a_{kl} and b_{kl} are polynomial coefficients of concentration and temperature, respectively. $F_T(b_{kl})$ and $F_C(a_{kl})$, respectively, represent polynomials describing the temperature concentration distribution. Thus, the number of basis functions is $N_{bf} = (m + 1)(m + 2)/2$, leading to $2N_{bf}$ unknown coefficients for both temperature and concentration. The integrated absorbance signal and its corresponding error for wavelength λ can be expressed as follows when the number of projection paths is designated as P_r, the total grid number is defined as Q, and the total number of wavelengths is specified as N_λ:

$$A_{\lambda,p} = \sum_{q=1}^{Q} L_{p,q}\cdot C_q\cdot S_\lambda(T_q)\cdot P = \sum_{q=1}^{Q} L_{p,q}\cdot F_C(a_{kl})_q\cdot S_\lambda\left[F_T(b_{kl})_q\right]\cdot P \tag{5}$$

$$Error = \sum_{\lambda,p}\left[(A_{\lambda,p})_{theroy} - (A_{\lambda,p})_{experiment}\right]^2 \tag{6}$$

Based on the physical a priori information, boundary constraints for the temperature and concentration fields are imposed for each grid cell, resulting in $2Q$ inequality constraints. With the projection paths fixed, the calculation of temperature and concentration can be reformulated in a matrix form as

$$T_q(x_q, y_q) = \sum_{k=0}^{m}\sum_{l=0}^{k} b_{k-l,l}\, x_q^{k-l}\, y_q^l = \sum_{h=1}^{N_{base}} b_h \Psi_{h,q} = (\Psi b)_q \tag{7}$$

where the basis function matrix Ψ is precomputed according to the grid partitioning. Thus, the constraints on temperature and concentration can be transformed into the form

$$\begin{aligned} T_{min} \leqq \Psi b \leqq T_{max} \\ C_{min} \leqq \Psi a \leqq C_{max} \end{aligned} \tag{8}$$

The overall algorithm is illustrated in Figure 1. It integrates multi-spectral measurements via TDLAS with physical a priori information, imposing upper and lower bounds on temperature and concentration for each grid cell. By solving the minimization of the error equations using a Sequential Quadratic Programming (SQP) algorithm, the bivariate polynomial coefficient matrices a and b in Equations (3) and (4) are determined, yielding the reconstructed temperature and concentration fields.

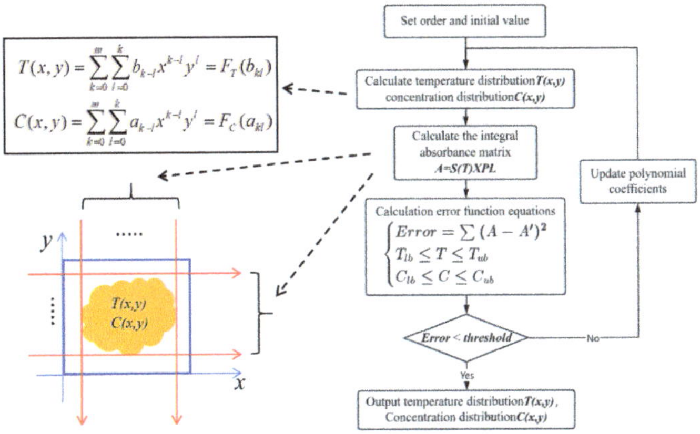

Figure 1. Process of the constrained polynomial model algorithm.

2.3. *Adaptive Polynomial Order Selection Algorithm*

A critical issue in the bivariate polynomial fitting of temperature and concentration fields is the selection of the polynomial order m. In practice, the precise distribution of the temperature and concentration fields is unknown. A polynomial with too low an order may be insufficient to describe a complex field, resulting in high reconstruction errors; conversely, a polynomial with too high an order increases computational time and the risk of converging to local minima due to an increased number of unknowns. In this study, an adaptive order selection algorithm based on error correction is developed. The algorithm flow is illustrated in Figure 2.

Initially, parameters are set, including the initial polynomial order (Initial_Order, typically set to 8), the maximum allowable order (Max_Order, typically 13), an initial residual value (Initial_Error, set to a large value), and a maximum number of iterations (N, set to 10). The Initial_Error serves to mark the current minimum residual, an Error_flag indicates the number of iterations X during which the error does not improve (typically $X = 3$), and an Order_flag tracks the consecutive increases in the polynomial order with no change in error (threshold typically set to 2). The algorithm begins with the lowest-order polynomial and iteratively computes the solution; if the residual does not improve for several consecutive iterations, the polynomial order is increased, and the parameters from the previous order are used as the initial values for further iterations. When the residual reaches a threshold, the maximum order is reached, or when the number of successive order increases exceeds the preset limit without error improvement, the algorithm terminates. This indicates that within the current order range, further increases do not significantly reduce the reconstruction error and only decrease computational efficiency. This method not only ensures sufficient reconstruction accuracy but also enhances efficiency and avoids errors associated with inappropriate order selection in unknown fields.

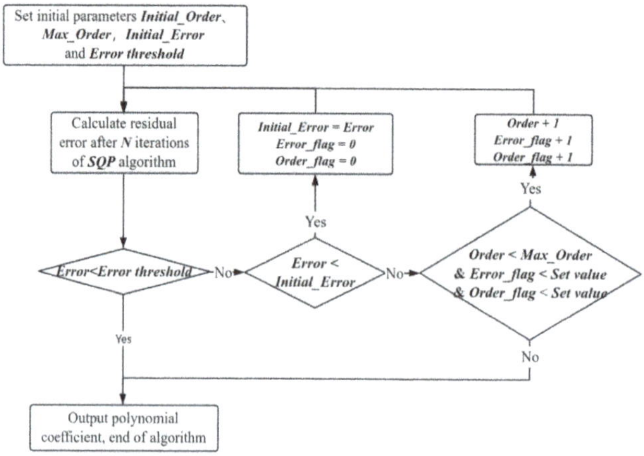

Figure 2. Process of adaptive polynomial order selection algorithm.

3. Results

3.1. Numerical Simulations

Numerical simulations are conducted to verify the performance of the proposed bivariate polynomial algorithm. Three different combustion field configurations are simulated. Figure 3a represents a single Gaussian peak distribution, Figure 3b represents a dual Gaussian peak distribution, and Figure 3c represents a mixed dual Gaussian distribution. For the single-peak and double-peak cases, the temperature ranges from 500 to 1500 K with a H_2O concentration between 0.05 and 0.12, while the mixed peak spans 500–1800 K with a H_2O concentration from 0.05 to 0.15. The measurement domain is discretized into a 30×30 grid. The side length of the simulated distribution is set at 6 cm with a spacing of 0.2 cm between beams.

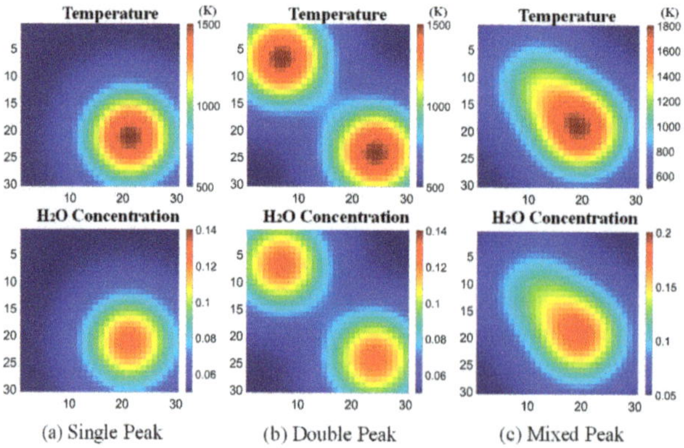

Figure 3. Three simulated fields: (**a**) single peak; (**b**) double peak; (**c**) mixed peak.

The simulation employs four projection angles (0°, 45°, 90°, and 135°) with 30 beams per angle, yielding a total of 120 laser paths as depicted in Figure 4 (red lines represent laser paths, and the black grid lines indicate the discretization). Four absorption wavelengths (1339 nm, 1343 nm, 1392 nm, and 1469 nm) are selected, and the integrated absorbance at these wavelengths is computed using absorption line parameters from the HITRAN database.

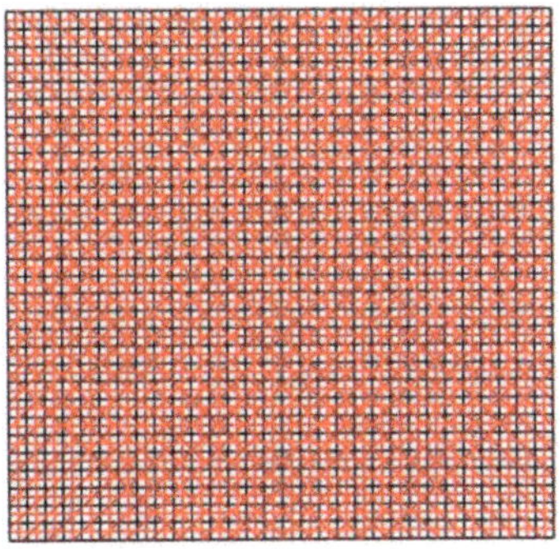

Figure 4. Laser path and grid.

The reconstruction errors for the temperature and concentration fields are defined as follows:

$$Error\_T = \frac{\sum_i^Q \left| T_i^{true} - T_i^{cal} \right|}{\sum_i^Q T_i^{true}}$$

$$Error\_C = \frac{\sum_j^Q \left| C_j^{true} - C_j^{cal} \right|}{\sum_j^Q C_j^{true}}$$

where Q is the total number of grid cells, T^{true} and C^{true} represent the simulated temperature and concentration values, and T^{cal} and C^{cal} are the corresponding reconstructed values.

3.1.1. Analysis of Polynomial Order Effects

The order of the bivariate polynomial affects the model's ability to accurately represent the surface; higher orders offer greater descriptive power but lead to increased computational time and risk of underdetermination due to a surge in unknown coefficients. In contrast, lower orders may fail to capture the complexity of the field, resulting in significant reconstruction errors.

For the single-peak, dual-peak, and mixed fields, the reconstruction error between the bivariate polynomial model and the simulated temperature field is computed for various polynomial orders, with the results shown in Figure 5. The results indicate that for lower orders (below 9), the error differences among different field types are significant, potentially reducing the model's generalization capability. When the order exceeds 12, the reconstruction errors across the three configurations become comparable, suggesting that a higher-order polynomial can universally describe diverse combustion fields. However, as shown by the increase in the number of polynomial coefficients with order, an excessively high order leads to a rapid increase in unknowns, which may cause underdetermined problems and significantly raise computational costs.

The proposed adaptive order adjustment algorithm based on error variation addresses the difficulty of pre-determining the polynomial order in unknown field reconstructions. Based on the performance and computational cost analysis, the polynomial order range is set between 8 and 13, ensuring applicability to most combustion fields. All subsequent simulations employ this adaptive strategy.

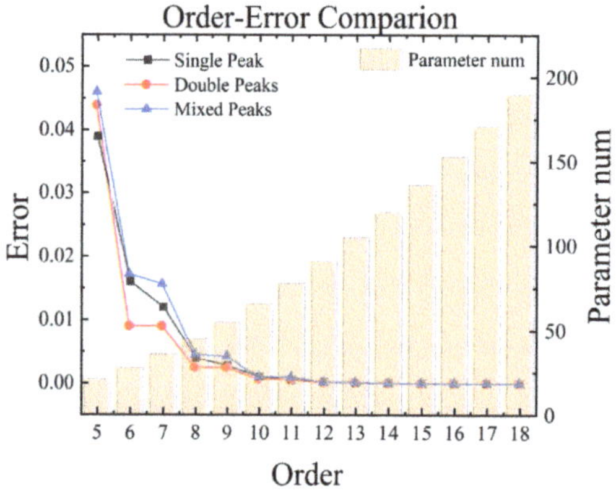

Figure 5. Error of different orders.

3.1.2. Simulation Results

Currently, two main types of algorithms are used for reconstructing two-dimensional combustion fields: linear tomography methods (e.g., ART) and nonlinear tomography methods. In the linear approach, ART is compared with both the non-constraint polynomial fitting (NCPF) and the proposed constrained polynomial fitting (CPF) algorithms. Derived from the CPF model, the NCPF algorithm implementation removes constraint conditions while retaining core polynomial fitting principles in its computational model. In these comparisons, the temperature constraint is set to [296 K, 2500 K] and the concentration constraint to [0.0001, 0.3]. The ART algorithm uses a four-wavelength Boltzmann plot method for temperature and concentration reconstruction. The reconstruction results for the three simulated configurations are presented in Figure 6 and Table 1.

Table 1. Reconstruction error of temperature and concentration distribution.

	(a) Single Peak		
	NCRF	CPF	ART
Temperature	13.3%	2%	7.4%
Concentration	47.1%	2%	14.6%
	(b) Double Peak		
	NCRF	CPF	ART
Temperature	6.2%	1.6%	3.1%
Concentration	29.5%	1.8%	7.9%
	(c) Mixed Peak		
	NCRF	CPF	ART
Temperature	2.9%	2%	3%
Concentration	11.5%	2.6%	7.9%

Figure 6a–c show that while the NCPF algorithm yields satisfactory results for a simple single-peak field, its performance degrades significantly for more complex dual-peak and mixed fields, resulting in configuration distortions. Moreover, the unconstrained method exhibits pronounced anomalies in the peripheral regions, which are attributable to the insufficient number of laser paths at the edges. These results indicate that, without physical a priori constraints, the NCPF algorithm cannot produce reliable reconstructions. In contrast, the CPF algorithm yields accurate reconstructions for all three configurations, with overall consistency between the reconstructed and simulated fields. The temperature

reconstruction errors for the three configurations are 2.03%, 1.58%, and 1.96%, respectively, demonstrating the method's versatility. According to the error values in Table 1, compared with the NCPF algorithm, the improved CPF method achieves increases in temperature accuracy of 84.9%, 74.4%, and 31.9% and concentration accuracy improvements of 95.8%, 94%, and 77.1%, respectively.

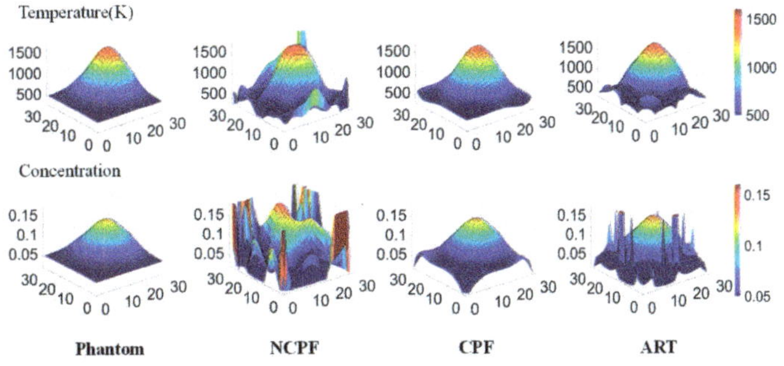

(**a**) Reconstruction results of single peak.

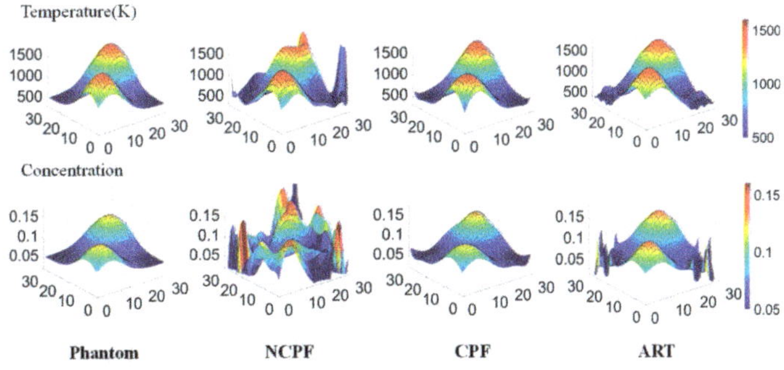

(**b**) Reconstruction results of double peak.

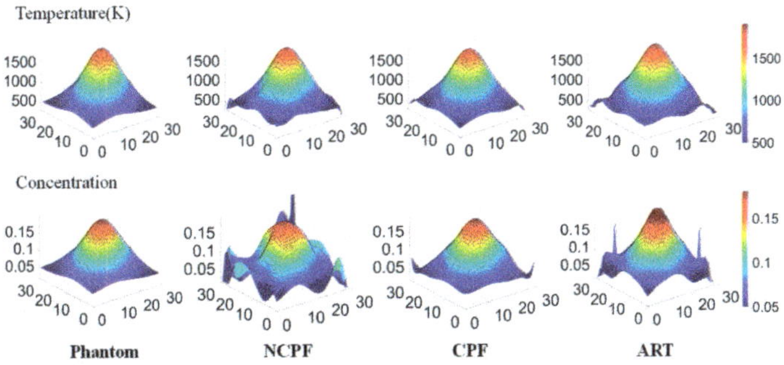

(**c**) Reconstruction results of mixed peak.

Figure 6. Reconstruction results of temperature and concentration distribution: (**a**) single peak; (**b**) double peak; (**c**) mixed peak.

Figure 7 presents the one-dimensional temperature profiles along a line parallel to the x-axis through the peak region. The CPF algorithm is shown to accurately capture the overall temperature trend: the high-temperature region (>1000 K) exhibits one-dimensional absolute mean differences of 55 K, 10.3 K, and 31.7 K, while the low-temperature region ($\leq$1000 K) shows mean differences of 19.5 K, 6.9 K, and 16.4 K. Compared to the traditional ART algorithm, the CPF algorithm consistently reduces reconstruction errors across all configurations, particularly in concentration, and maintains overall errors below 3%. Furthermore, unlike the NCPF and ART methods, which display marked peripheral anomalies, the CPF method effectively suppresses these anomalies, yielding a smoother and more accurate reconstruction.

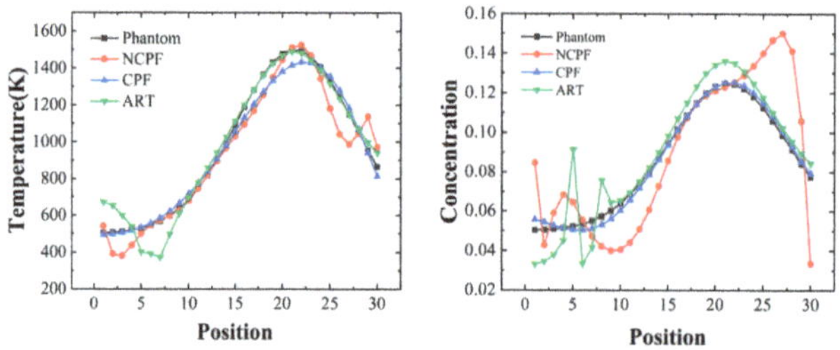

(**a**) single peak.

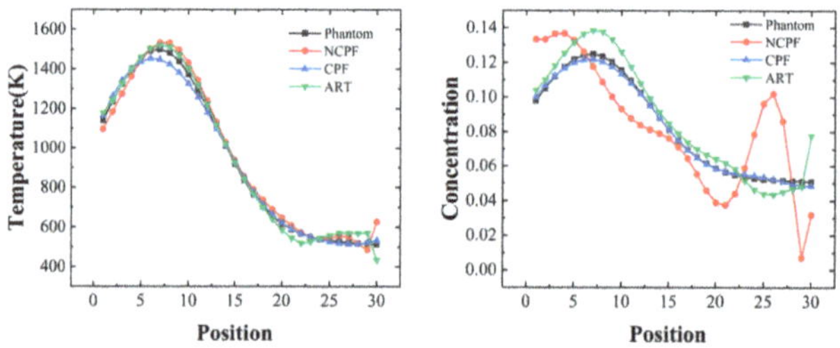

(**b**) double peak.

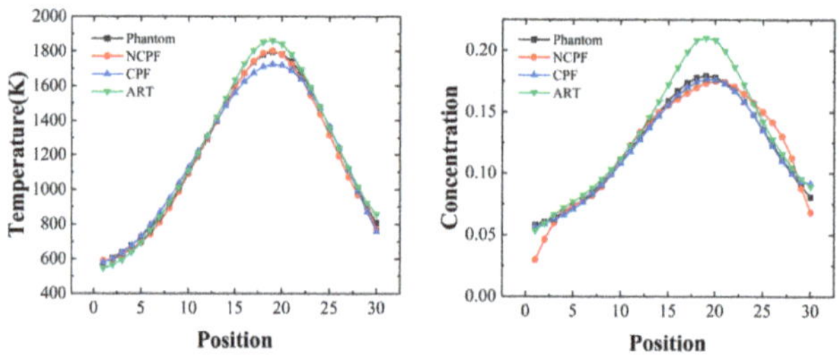

(**c**) mixed peak.

Figure 7. Temperature and concentration distribution curve with peak point parallel to X axis: (**a**) comparison results of single peak; (**b**) comparison results of double peak; (**c**) comparison results of mixed peak.

3.2. Effect of Noise on Reconstruction Accuracy

The superior performance of the CPF algorithm is further evidenced by its contrast with the NCPF method, which exhibits significant errors and distorted concentration reconstructions. To assess robustness, Gaussian white noise levels ranging from 0.5% to 5% are added to the integrated absorbance matrix of the three simulated distributions. For each noise level, 30 simulation trials are performed to mitigate random fluctuations, and both the CPF and ART algorithms are applied. The average reconstruction errors for the three configurations are shown in Figure 8, with error bars representing the standard deviation over 30 trials. Although the CPF algorithm's error increases more rapidly with noise compared to ART, the overall error remains low. Under 5% Gaussian white noise, the temperature reconstruction errors for the three configurations are 8.48%, 7.02%, and 6.01%. Notably, ART exhibits smaller variations with noise, indicating higher stability, while the CPF algorithm's stability requires further improvement.

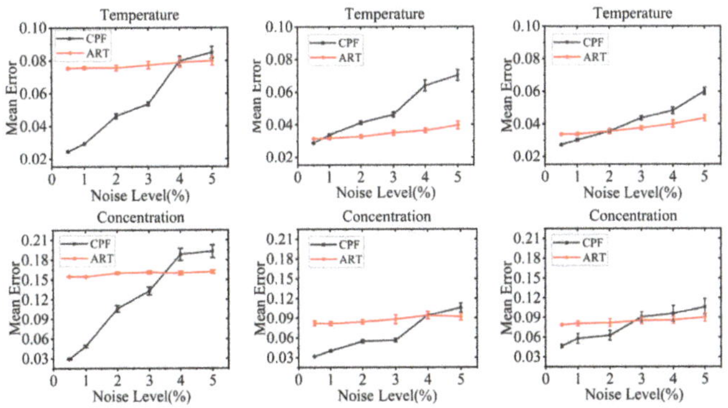

(a) Single Peak (b) Double Peak (c) Mixed Peak

Figure 8. Reconstruction error under 1–5% Gaussian noise: (**a**) reconstruction error of single peak; (**b**) reconstruction error of double peak; (**c**) reconstruction error of mixed peak.

Table 2 details the average error variations for both algorithms under different noise levels. In terms of concentration reconstruction, for the single Gaussian peak model, the CPF algorithm achieves a concentration error of 2.94% under 0.5% noise—an 82.1% reduction compared with ART's error of 15.47%. For the dual-peak and mixed distributions, the CPF concentration errors remain lower than those of ART when the noise is below 4%; for the mixed distribution, however, under 5% noise, the CPF error (10.62%) is 18.4% higher than ART (8.97%), although under low noise (0.5%), CPF still outperforms ART. In temperature reconstruction, CPF shows superior accuracy under low-noise conditions (noise level $\leq$ 2%); for instance, in the single-peak model under 0.5% noise, the error is 2.45%, only 32.6% of ART's 7.51%. Similarly, in the mixed distribution under 0.5% noise, CPF achieves an error of 2.73%, 18.8% lower than ART's 3.36%. Despite the more rapid error growth under high noise (noise level $\geq$ 4%), the overall performance and global advantage in concentration reconstruction highlight the CPF algorithm's adaptability to complex combustion fields, providing a reliable theoretical basis for high-precision diagnostics.

Table 2. Reconstruction error of different configurations.

(a) Single Peak							
Noise Level		0.5%	1%	2%	3%	4%	5%
CPF	Temperature	2.45%	2.91%	4.61%	5.35%	7.96%	8.48%
	Concentration	2.94%	4.85%	10.58%	13.3%	18.82%	19.28%
ART	Temperature	7.51%	7.55%	7.54%	7.71%	7.88%	7.97%
	Concentration	15.47%	15.43%	15.98%	16.08%	15.97%	16.16%
(b) Double Peak							
Noise Level		0.5%	1%	2%	3%	4%	5%
CPF	Temperature	2.86%	3.36%	4.11%	4.61%	6.38%	7.02%
	Concentration	3.16%	4.01%	5.45%	5.61%	9.42%	10.6%
ART	Temperature	3.12%	3.16%	3.26%	3.48%	3.63%	3.94%
	Concentration	8.18%	8.13%	8.39%	8.77%	9.38%	9.15%
(c) Mixed Peak							
Noise Level		0.5%	1%	2%	3%	4%	5%
CPF	Temperature	2.73%	3.02%	3.55%	4.36%	4.82%	6.01%
	Concentration	4.64%	5.79%	6.24%	9.05%	9.58%	10.62%
ART	Temperature	3.36%	3.38%	3.55%	3.75%	3.99%	4.34%
	Concentration	7.85%	8.05%	8.17%	8.51%	8.57%	8.97%

3.3. Experimental Verification

Following the numerical simulations, the proposed algorithm was further validated using a standard McKenna flat-flame burner in a laboratory setting to assess its feasibility in practical combustion fields. The TDLAS system and a thermocouple measurement system were used concurrently under the same operating conditions, with the measurement plane located 50 mm above the burner surface. As illustrated in Figure 9, the TDLAS scanning range extended from the burner center to its periphery, with a 4 mm shift per measurement (totaling 12 positions). The thermocouple measurement path corresponded with the TDLAS scanning path. In the figure, the red lines represent the actual laser paths, the black dashed lines indicate the reconstruction grid (4 mm grid size), and the circular region denotes the actual combustion zone (6 cm in diameter). Green dots represent thermocouple measurement points, spaced consistently with the laser paths. Due to the axisymmetric distribution of the flat-flame burner, single-angle measurements were transformed into equivalent data for four angles (0°, 45°, 90°, and 135°) using an equivalent optical path method. Background absorption interference was removed by proportionally subtracting background signals from the peripheral measurements. The final reconstruction resolution was 17 × 17, with each grid measuring 4 mm × 4 mm, yielding a square reconstruction area of 68 mm per side that is centered on the burner and covers the entire combustion zone.

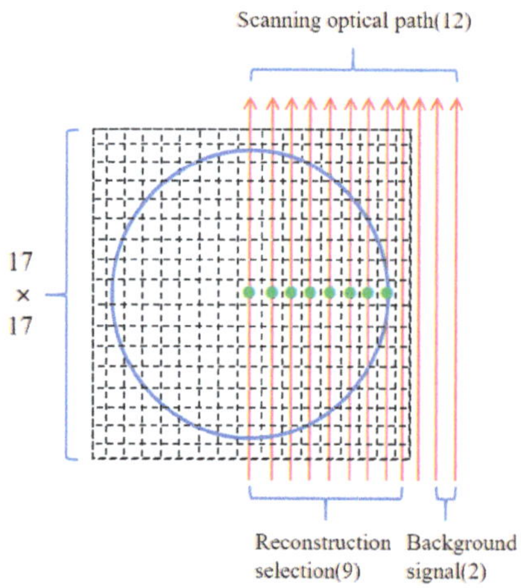

Figure 9. Laser path of the experiment.

The experimental setup of the TDLAS system is depicted in Figure 10. The TDLAS system employs five DFB lasers operating at discrete wavelengths of 1339 nm, 1343 nm, 1392 nm, 1398 nm, and 1469 nm, with time-division multiplexing synchronization. Four characteristic absorption lines (1339 nm, 1343 nm, 1392 nm, and 1469 nm) were algorithmically selected for spectral analysis. A beam splitter configuration established dual measurement paths: a probing beam traversing the combustion field and a reference beam propagating through a temperature-stabilized etalon. Both optical signals were acquired by matched photodetectors, with subsequent signal conditioning comprising low-noise amplification prior to digitization via a data acquisition system. The acquired spectral datasets were stored for subsequent multivariate analysis. A translation stage controlled the laser scanning with a step size of 4 mm. To mitigate the effects of flame pulsation during measurement, 4000 cycles of data were acquired at each position (approximately 1 min of acquisition per position) and averaged. The equivalence ratio of the flat-flame burner was maintained at 0.8 in continuous combustion until all data were collected. The reconstruction results obtained from the algorithm are shown in Figure 11.

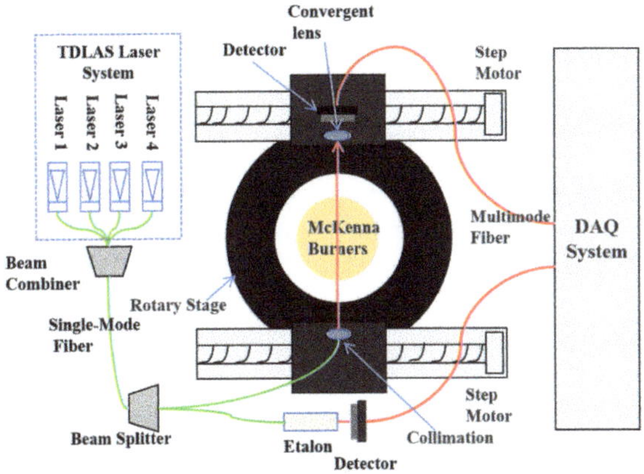

Figure 10. TDLAS measurement system.

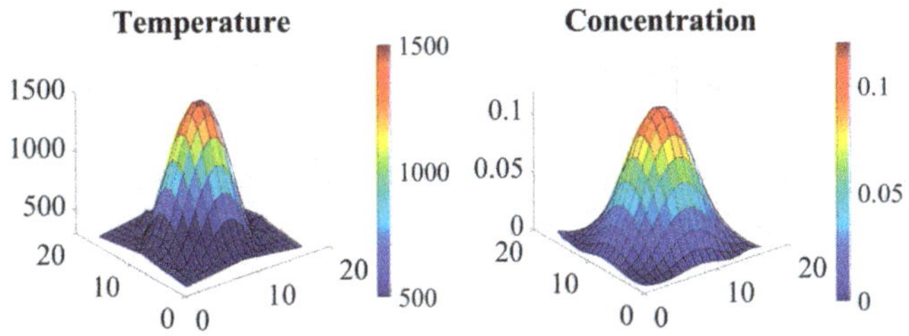

Figure 11. Reconstruction results of experimental temperature and concentration distribution.

Figure 12 compares the reconstructed temperatures with thermocouple measurements. The experimental results demonstrate excellent agreement in the high-temperature central region, with a temperature difference of only 6 K at the center (position = 10), indicating high accuracy in the central region. The overall temperature trend in the reconstructed results closely follows that of the thermocouple measurements. In the high-temperature region (>1000 K), the average temperature difference was only 33 K, whereas in the low-temperature region ($\leq$1000 K), the average difference reached 188 K. The two-dimensional reconstruction also reveals anomalously high temperatures at the periphery, indicating partial distortion in low-temperature areas.

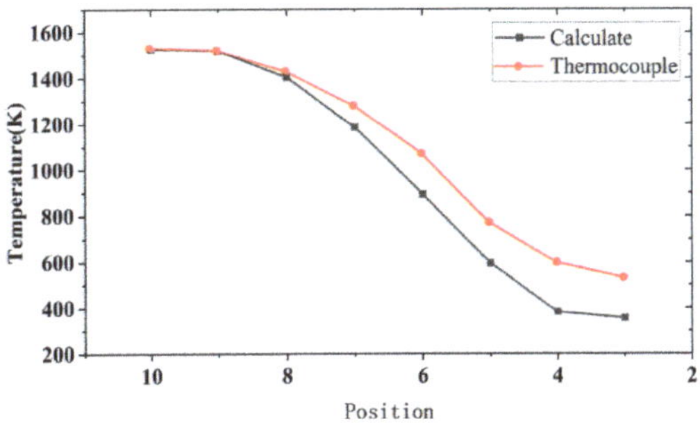

Figure 12. Algorithm reconstruction results and thermocouple measurement results.

Figure 13 presents the raw measurement signals, with the black line representing the central region and the red line the periphery. It can be observed that the absorption in the peripheral region is significantly weaker compared to the center. Notably, the absorption peaks for the 1343 nm and 1339 nm lines are almost indiscernible, and those for the 1392 nm and 1469 nm lines are markedly diminished. As indicated by the one-dimensional comparisons in Figure 7, although the algorithm describes the low-temperature region reasonably well, its performance degrades under high noise conditions. This suggests that the pronounced deviations in the low-temperature region are due to a reduced effective optical path and lower signal-to-noise ratio (SNR) as the temperature decreases toward the burner edge, resulting in diminished measurement sensitivity.

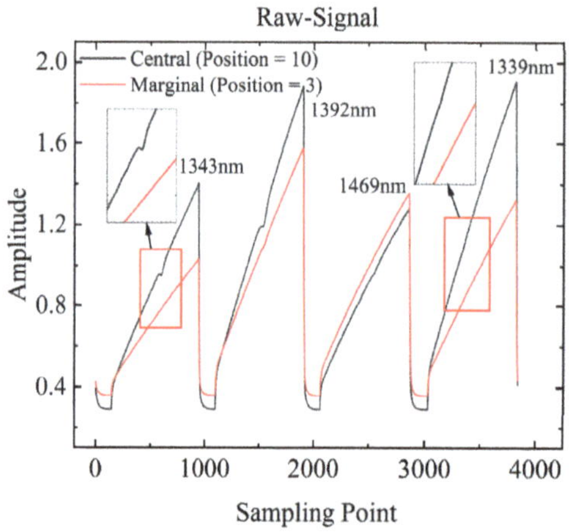

Figure 13. Original signals in the center and the edge of the experiment.

4. Conclusions

Based on the polynomial CT-TDLAS algorithm, this study proposes a nonlinear constrained polynomial fitting algorithm. By employing a single-objective optimization method for the integrated absorbance under constraint conditions—rather than the traditional multi-objective approach—and incorporating an adaptive order selection strategy, the proposed method addresses the challenge of determining the appropriate polynomial order for unknown fields. Comparisons with the traditional ART algorithm demonstrate that, for three distinct distribution configurations, the proposed algorithm achieves reconstruction errors below 3%, outperforming ART. Moreover, when Gaussian white noise levels of 0–5% are introduced, the overall error remains low, underscoring the method's robustness. However, compared with ART, the CPF algorithm exhibits greater error variability, indicating that its stability may still be improved.

Experimental validation on a flat-flame burner shows that in high-temperature regions—especially at the center—the reconstruction error is extremely low, with a central difference of only 6 K (approximately 0.3%). The overall trend of the reconstructed temperature field is consistent with that of the thermocouple measurements. Nevertheless, discrepancies remain in the low-temperature regions, which are attributed to reduced SNRs resulting from weaker absorbance signals. Future work will further investigate the impact of absorption line selection on the algorithm to enhance its performance in low-temperature regions and improve reconstruction stability.

Author Contributions: Conceptualization, C.C.; methodology, C.C.; software, C.C. and Z.X.; resources, Z.X.; data curation, A.H. and R.N.; writing—original draft preparation, C.C. and S.A.;

writing—review and editing, C.C. and D.S.; visualization, C.C. and T.J.; supervision, Z.X.; project administration, Z.X.; funding acquisition, Z.X. All authors have read and agreed to the published version of the manuscript.

Funding: The authors declare that financial support was received for the research, authorship, and/or publication of this article. This research was funded by the National Key R&D Program of China (Grant No. 2023YFF0716400), the Youth Innovation Promotion Association of Chinese Academy of Sciences (Grant No. 2022451), and the Stable Support for Institutes-Research (Grant No. 0200850008).

Data Availability Statement: The original contributions presented in the study are included in the article further inquiries can be directed to the corresponding author.

Conflicts of Interest: The authors declare no conflicts of interest.

Abbreviations

The following abbreviations are used in this manuscript:

CT-TDLAS	Computed Tomography–Tunable Diode Laser Absorption Spectroscopy
ART	Algebraic Reconstruction Technique
RBF	Radial Basis Function
SA	Simulated Annealing
CPF	Constrained Polynomial Fitting
SQP	Sequential Quadratic Programming
NCPF	Non-Constraint Polynomial Fitting

References

1. Rieker, G.; Jeffries, J.; Hanson, R.; Mathur, T.; Gruber, M. Diode laser-based detection of combustor instabilities with application to a scramjet engine. *Proc. Combust. Inst.* **2009**, *32*, 831–838. [CrossRef]
2. Spearrin, R.M.; Goldenstein, C.S.; Schultz, I.A.; Jeffries, J.B.; Hanson, R.K. Simultaneous sensing of temperature, CO, and CO_2 in a scramjet combustor using quantum cascade laser absorption spectroscopy. *Appl. Phys. B* **2014**, *117*, 689–698. [CrossRef]
3. Schultz, I.A.; Goldenstein, C.S.; Spearrin, R.M.; Jeffries, J.B.; Hanson, R.K.; Rockwell, R.D.; Goyne, C.P. Multispecies midinfrared absorption measurements in a hydrocarbonfueled scramjet combustor. *J. Propul. Power* **2014**, *30*, 1595–1604. [CrossRef]
4. Schultz, I.A.; Goldenstein, C.S.; Jeffries, J.B.; Hanson, R.K.; Rockwell, R.D.; Goyne, C.P. Diode laser absorption sensor for combustion progress in a model scramjet. *J. Propuls. Power* **2014**, *30*, 550–557. [CrossRef]
5. Schultz, I.; Goldenstein, C.; Jeffries, J.; Hanson, R.; Rockwell, R.; Goyne, C. TDL absorption sensor for in situ determination of combustion progress in scramjet ground testing. In Proceedings of the 28th Aerodynamic Measurement Technology, Ground Testing, and Flight Testing Conference including the Aerospace T&E Days Forum, New Orleans, LA, USA, 25–28 June 2012.
6. Chang, L.S.; Strand, C.L.; Jeffries, J.B.; Hanson, R.K.; Diskin, G.S.; Gaffney, R.L.; Capriotti, D.P. Supersonic mass-flux measurements via tunable diode laser absorption and nonuniform flow modeling. *AIAA J.* **2011**, *49*, 2783–2791. [CrossRef]
7. Busa, K.; Bryner, E.; McDaniel, J.; Goyne, C.; Smith, C.; Diskin, G. Demonstration of capability of water flux measurement in a scramjet combustor using tunable diode laser absorption tomography and stereoscopic PIV. In Proceedings of the 49th AIAA Aerospace Sciences Meeting Including the New Horizons Forum and Aerospace Exposition, Orlando, FL, USA, 4–7 January 2011.
8. Fu, P.; Hou, L.; Chao, X.; Wang, Z. Time-resolved multi-parameter detection in a kerosene-fueled combustor using tunable diode laser absorption spectroscopy. *Exp. Therm. Fluid Sci.* **2023**, *148*, 110978. [CrossRef]
9. Bao, Y.; Zhang, R.; Enemali, G.; Cao, Z.; Zhou, B.; McCann, H.; Liu, C. Relative entropy regularized TDLAS tomography for robust temperature imaging. *IEEE Trans. Instrum. Meas.* **2020**, *70*, 4501909. [CrossRef]
10. Gao, X.; Cao, Z.; Tian, Y.; Xu, L. RBF-based reconstruction method for tomographic imaging of temperature and water vapor concentration in flames. In Proceedings of the 2021 IEEE International Conference on Imaging Systems and Techniques (IST), New York, NY, USA, 24 August 2021; IEEE: Piscataway, NJ, USA, 2021; pp. 1–6.
11. Busa-Rice, K.M.; Brown, M.S.; Nadir, Z.; Bouman, C.A.; Comer, M.L. Implementation of Model-Based Iterative Reconstruction Approach to Tomographic Tunable Diode Laser Absorption Spectroscopy. In Proceedings of the 2018 AIAA Aerospace Sciences Meeting, Kissimmee, FL, USA, 8–12 January 2018; Volume 1361.
12. Ma, L.; Cai, W. Numerical investigation of hyperspectral tomography for simultaneous temperature and concentration imaging. *Appl. Opt.* **2008**, *47*, 3751–3759. [CrossRef]
13. Yi-Peng, Z.; Zhen-Yu, X.; An, H.; Ai, S.-M.; Xia, H.-H.; Kan, R.-F. Distribution reconstruction of non-uniform combustion field based on improved simulated annealing algorithm. *Acta Phys. Sin.* **2021**, *70*, 134205.

14. Huang, J.; Liu, H.; Dai, J.; Cai, W. Reconstruction for limited-data nonlinear tomographic absorption spectroscopy via deep learning. *J. Quant. Spectrosc. Radiat. Transf.* **2018**, *218*, 187–193. [CrossRef]
15. Yu, T.; Cai, W.; Liu, Y. Rapid tomographic reconstruction based on machine learning for time-resolved combustion diagnostics. *Rev. Sci. Instruments* **2018**, *89*, 043101. [CrossRef] [PubMed]
16. Si, J.; Li, G.; Cheng, Y.; Zhang, R.; Enemali, G.; Liu, C. Hierarchical temperature imaging using pseudoinversed convolutional neural network aided TDLAS tomography. *IEEE Trans. Instrum. Meas.* **2021**, *70*, 4506711. [CrossRef]
17. Wang, Z.; Zhou, W.; Kamimoto, T.; Deguchi, Y.; Yan, J.; Yao, S.; Girase, K.; Jeon, M.-G.; Kidoguchi, Y.; Nada, Y. Two-dimensional temperature measurement in a high-temperature and high-pressure combustor using computed tomography tunable diode laser absorption spectroscopy (CT-TDLAS) with a wide-scanning laser at 1335–1375 nm. *Appl. Spectrosc.* **2020**, *74*, 210–222. [CrossRef] [PubMed]
18. Allen, M.G. Diode laser absorption sensors for gas-dynamic and combustion flows. *Meas. Sci. Technol.* **1998**, *9*, 545. [CrossRef] [PubMed]
19. Kan, R.F.; Xia, H.H.; Xu, Z.Y.; Yao, L.; Ruan, J.; Fan, X. Research and progress of flow field diagnosis based on laser absorption spectroscopy. *Chin. J. Lasers* **2018**, *45*, 0911005.
20. Cai, W.; Kaminski, C.F. Tomographic absorption spectroscopy for the study of gas dynamics and reactive flows. *Prog. Energy Combust. Sci.* **2017**, *59*, 1–31. [CrossRef]
21. Jeon, M.G.; Deguchi, Y.; Kamimoto, T.; Doh, D.H.; Cho, G.R. Performances of new reconstruction algorithms for CT-TDLAS (computer tomography-tunable diode laser absorption spectroscopy). *Appl. Therm. Eng.* **2017**, *115*, 1148–1160. [CrossRef]

photonics

MDPI

Communication

A Single-Channel Correction Method for Spectral Responsivity Differences in Detector Arrays

Yilun Cheng [1,2], Fengfu Tan [1,3,*], Gangyu Wang [1,2], Yang Li [1,4], Laian Qin [1,3], Feng He [1,3] and Zaihong Hou [1,3]

1 Key Laboratory of Atmospheric Optics, Anhui Institute of Optics and Fine Mechanics, Hefei Institutes of Physical Science, Chinese Academy of Sciences, Hefei 230031, China; ylcheng@aiofm.ac.cn (Y.C.); gywang@aiofm.ac.cn (G.W.); ly2020@mail.ustc.edu.cn (Y.L.); laqin@aiofm.ac.cn (L.Q.); fhe@aiofm.ac.cn (F.H.); zhhou@aiofm.ac.cn (Z.H.)
2 Nanhu Laser Laboratory, National University of Defense Technology, Changsha 410073, China
3 State Key Laboratory of Laser Interaction with Matter, Anhui Institute of Optics and Fine Mechanics, Hefei Institutes of Physical Science, Chinese Academy of Sciences, Hefei 230031, China
4 Science Island Branch of Graduate School, University of Science and Technology of China, Hefei 230026, China
* Correspondence: tff@aiofm.ac.cn

Abstract: The spectral responsivity of photodetectors exhibits significant variations across different wavelengths. Such variations can induce substantial errors when large-scale detector array modules are employed for the measurement of laser spot parameters. In this regard, a single-channel data correction methodology is proposed herein to mitigate the spectral responsivity discrepancies within large-scale detector arrays. Specifically, the single-channel incident laser within the detector array is bifurcated and irradiated onto the detector with a coated window mirror and the detector at the original corresponding position, respectively. Subsequently, the correction coefficient is computed based on the single-channel data, thereby effectuating the correction of spectral response differences within the large-scale detection array. Through this approach, the measurement error resulting from the spectral responsivity differences in the detection array measurement system is diminished to less than 2%. Notably, this method is applicable to large-scale detection arrays and is not circumscribed to the domain of laser parameter measurement.

Keywords: spectral responsivity; single-channel correction; detector arrays

1. Problem Retrospect

The detector array method has witnessed increasingly extensive application in the realm of laser parameter measurement technology. This is primarily attributed to its ability to directly measure the light spot, the wide dynamic range of the detector, the high sampling frequency of the system, and the favorable real-time performance. It can also fulfill various test environment requirements [1–3]. The hardware architecture is illustrated in Figure 1. In the detector array method, photodetectors are utilized to convert optical signals into electrical signals. Subsequently, the laser spot parameters at the target end are computed by the data processing system, thereby facilitating the evaluation of the performance of the laser emission system. Alternatively, the measured spot parameters can be employed to investigate the characteristics of high-energy laser atmospheric transmission [4–6]. With the incessant advancement of laser technology, there is a growing demand for laser beams with high output power and excellent beam quality, which are achieved through the modular synthesis of multiple lasers [7–9]. Generally, laser beam combining methods can be categorized into two types: coherent combining technology and incoherent combining technology. The incoherent beam combining technology mainly encompasses laser fiber combining, spatial array combining, spectral combining, and polarization combining [10,11]. With the development of key technologies such as heat resistance and high diffraction efficiency in core devices like multilayer dielectric gratings, spectral synthesis technology has manifested its substantial development potential [12,13]. The basic structure of the spectral synthesis method of the laser beam is presented in Figure 2, the curves of different colors in the diffraction grattings and fiber array represent light beams of different wavelengths, and the out beam represents the laser after the combination. Owing to the extensive spectral range of the synthesized spectrum, photodetectors exhibit varying spectral responsivity across different bands, which inevitably gives rise to significant errors in the measurement data [14,15]. Hence, in this study, the impact of differences in the detector's spectral responsivity on the measurement accuracy is analyzed. Based on this analysis, a method for correcting the spectral response differences in the large-scale detector array measurement system using single-channel data is proposed.

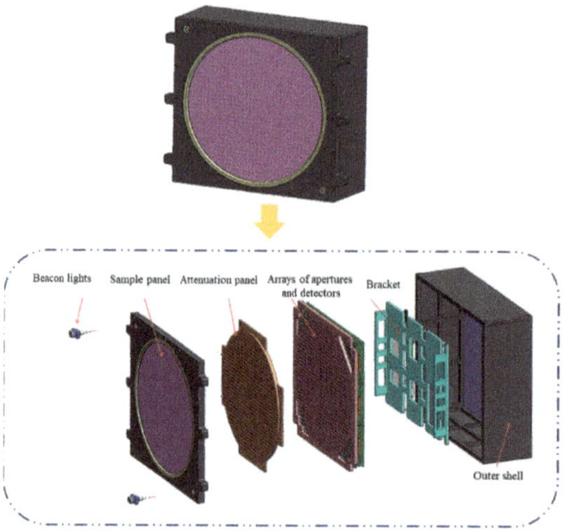

Figure 1. Structure diagram of detector array method.

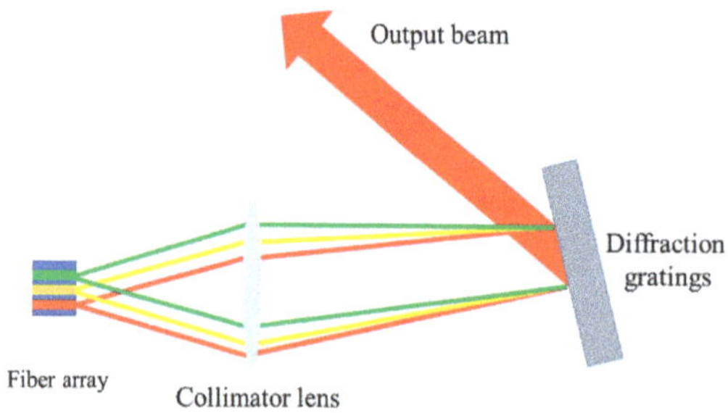

Figure 2. Basic structure of spectral synthesis.

The InGaAs near-infrared detector G12180-010A, manufactured by Hamamatsu Optoelectronics Co., Ltd. in Hamamatsu, Japan, is taken as a reference. Its responsivity curve is depicted in Figure 3. As illustrated in Figure 3, within the wavelength range from 1030 nm to 1080 nm of the incident laser, the spectral responsivity of the detector exhibits a variation of approximately 6%. This implies that when the measurement system employs a single-wavelength laser for power density calibration, the maximum error of the system can reach around 6%. Hence, in the process of measuring the parameters of the intense laser spot in conjunction with the spectral synthesis method, it is imperative to rectify the differences in the spectral response of the detector across different wavelengths.

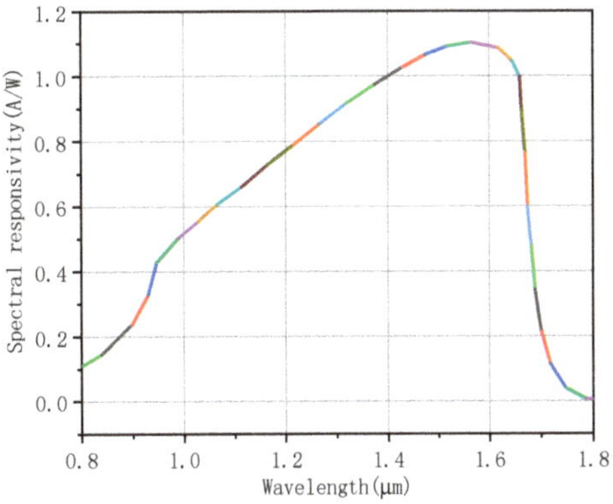

Figure 3. Spectral responsivity curve of the detector.

In the measurement system based on the detector array method, it is initially a requisite to quantitatively calibrate the system using a laser with a stable working mode and steady output power [16,17]. The laser with stable output is directed onto each channel of the detector array, and the correlation between the output current value of each channel of the detector and the power density of the detector sampling hole is computed. Nevertheless, the traditional measurement system fails to take into account the issues stemming from the expansion of the wavelength range of the incident beam. This primarily arises from two factors. Firstly, the wavelength range of the laser beam emitted by the conventional laser emission system is relatively narrow, exerting minimal impact on the measurement system. Secondly, it is challenging to identify a laser possessing a suitable wavelength, stable working mode, and steady output power for calibrating the measurement system. With regard to the error induced by the difference in the spectral response of the detector [18,19], the quantum efficiency of the detector can be pre-measured by a quantum efficiency measuring instrument, and the spectral intensity of the incident laser can be monitored in real-time by a spectrometer. Subsequently, the quantum efficiency response values and the spectral intensities of each band are weighted and aggregated to derive the power conversion coefficient, which is capable of nullifying the difference in quantum efficiency. However, several significant problems emerge when employing a spectrometer to measure the incident laser spectrum in real-time for correcting the total power of the laser. Firstly, the installation location of the spectrometer dictates whether it can ensure the reception of laser irradiation and furnish the spectral response data of the incident laser. Consequently, it might be necessary to augment the number of spectrometer probes on the target surface. Additionally, under different laser power irradiations, it remains to be ascertained whether the dynamic range of the spectrometer can guarantee that its output response falls within its operational limits. Thirdly, since the spectrometer is an independent product, the data output therefrom require secondary development to be seamlessly integrated with the data processing module of the measurement system.

In summary, in accordance with the research findings presented above, this study employs the window mirror coating technique with the aim of mitigating the spectral response variation of the detector, thereby achieving precise quantification of the spectral synthesis laser spot parameters.

2. Methods

The underlying principle of the window mirror coating methodology hinges upon the design of a filter possessing varying transmittance within a specific spectral band, which is predicated on the spectral response curve of the photodetector. Through the efficacious modulation of the light intensity across diverse bands, the output response of the photodetector can ultimately be rendered consistent under laser irradiation of disparate wavelengths. In response to the issue of utilizing coated window mirrors to nullify the spectral responsivity disparity, our research cohort carried out an exhaustive investigation. The film simulation devised for InGaAs photodetectors is illustrated in Figure 3. As depicted in Figure 4, it is evident that in the wavelength range spanning from 1000 nm to 1100 nm, in the absence of the coated window mirror, the quantum efficiency difference approximates 19.6%. Nevertheless, upon the introduction of the coated window mirror into the system, the spectral responsivity difference of the detector within this band is diminished by over 2%. Hence, the aforementioned data comprehensively substantiate that the coating methodology can efficaciously mitigate the spectral response disparity of the detector.

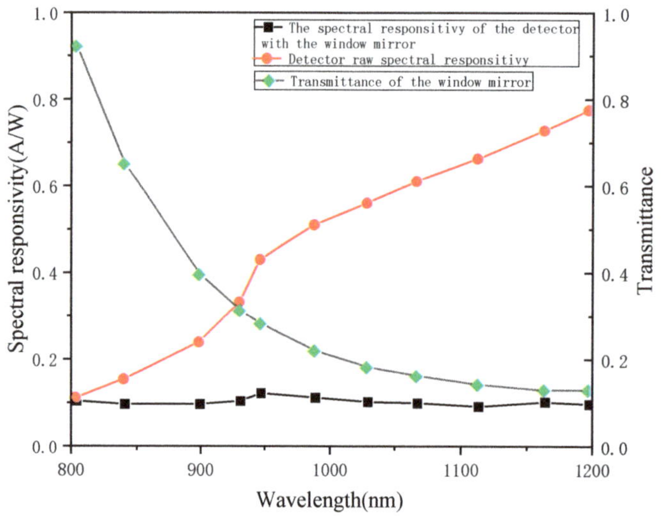

Figure 4. The spectral responsivity curve of the system after adding a coated window mirror.

Upon the application of the aforementioned method within the detector array measurement system, it is observed that the quantity of detectors present on the target surface is notably large. Under long-distance transmission scenarios, the target size of the detector array typically exceeds 30 cm × 30 cm. Consequently, in an attempt to nullify the spectral response difference of the detectors by installing a coated window mirror at the front end of the detectors, either a considerably large window mirror is necessitated or a substantial number of small-sized lenses are required. This invariably leads to a significant escalation in both the development cost of the measurement system and the intricacy of its structural configuration. Hence, this research proposes a methodology to mitigate the quantum efficiency difference by exploiting the single channel of the detector. The schematic diagram illustrating the structure of eliminating the quantum efficiency difference within the detector array measurement system via a single channel and a coated window mirror is presented in Figure 5. Within the central region of the detector array measurement system, a single-channel detector is chosen, and a beam splitter fiber module is installed at the rear end of the attenuation homogenization system. In this manner, one laser beam is directed towards the detector at its original corresponding position, while the other laser beam is irradiated onto the coated window mirror and subsequently impinges upon the detector after traversing the window mirror.

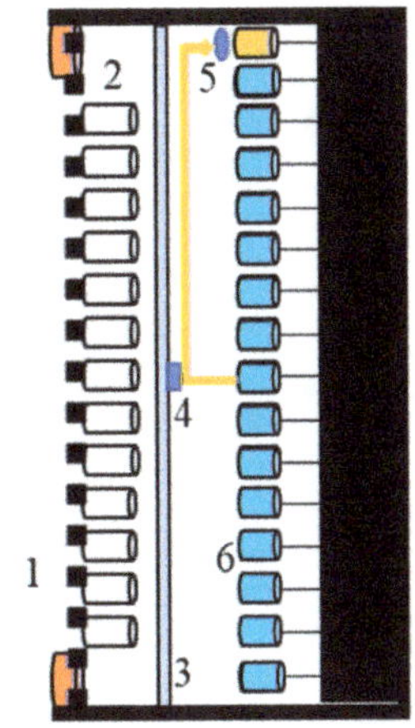

1. Sample panel; 2. Sample the fiber; 3. Beam homogenization system; 4. Spectroscopic fiber optic fiber module;5. Coated window mirror; 6. Detector array.

Figure 5. Single-channel spectral responsivity difference elimination system.

Essential Macleod software was used to simulate the design of the window mirror film layer. The characteristics of the film layer depended on the base material, the number of film layers, the thickness of the film layer, and the material of the film layer. The quantum efficiency difference was reduced by constantly changing the properties of the optical film layer in the Essential Macleod simulation software. The specific process is as follows:

Firstly, the quantum efficiency of the detector is measured by the quantum efficiency measuring instrument, and the spectral response data of the detector in the whole band are obtained.

Second, the initial reference wavelength, film thickness, film layer number, substrate material, and film layer material are set to obtain the corresponding film layer.

Finally, through analysis and comparison, the reference wavelength, film thickness, film layer number, substrate material, and film layer material are adjusted until the required film layer is suitable for the detector's working requirements.

In addition, the coating used in the simulation in this paper contains three parts, namely Na_3AlF_6, ZnS, and Al_2O_3.

3. Results

Within a specific wavelength range, the quantum efficiency of the detector spans from SR_1 to SR_n. Upon the installation of a coated window mirror at the front end of the detector, the quantum efficiency within this wavelength range alters to SR_m. Given that the measurement system is required to quantitatively compute the laser power parameters based on the incident optical signal, calibration against a standard light source is imperative during the development of the measurement system. This facilitates the acquisition of the power conversion coefficient K_{bd}, which represents the relationship between the output current value of the measurement system and the incident laser power. A single-wavelength narrow-linewidth light source, characterized by a stable laser working mode and stable power, is selected for this calibration purpose. The output wavelength of the calibrated light source is denoted as λ_{bd}, its power as P_{bd}, and the quantum efficiency of the detector at this wavelength is recorded as QE_{bd}. Subsequently, the laser power conversion coefficient of the channel devoid of the window mirror detector is computed as $K_{00} = 1/SR_{bd}$, while that of the channel equipped with the coated window mirror is calculated as $K_{01} = 1/SR_m$. Generally speaking, the wavelength range of the light source employed for calibration ought to align with that of the actual measured laser emission system. However, for certain specific wavelengths, a stable calibration light source is notably absent. Consequently, in the actual calibration process, single-wavelength calibration is typically adopted, and the calibration process after traversing a narrowband filter is illustrated in Figure 6; due to the attenuation system and diaphragm, the laser light impinges upon the beam splitter. Here, 50% of the laser is directed towards the detector array target, while the remaining 50% is incident upon the power meter. On the one hand, the power conversion coefficient of the detector can be derived from the measured value of the power meter and the output current value of the single channel of the detector array, namely the aforementioned K_{bd}. On the other hand, by stabilizing the illumination of the light source, the uniformity of all detector units on the target surface can be calibrated, thereby ensuring a consistent output response.

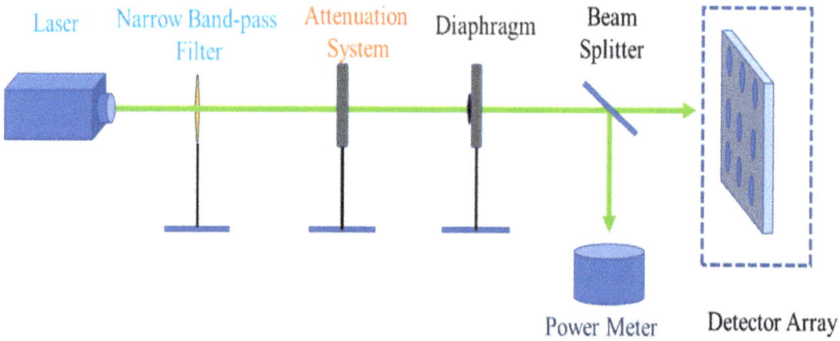

Figure 6. Detector array target calibration system.

During the measurement of the laser spot pertaining to spectral synthesis, the incident laser, upon traversing the sampling hole, is bifurcated into two beams. These two beams are then directed towards the detector equipped with a coated window mirror and the detector devoid of such a coating, respectively. Suppose the spectral range of the incident laser spans from λ_1 to λ_n; the original spectral responsivity of the detector within this band is denoted as SR_1 to SR_n, and the intensity within each band is P_1 to P_n. Subsequently, the output power P_{00} of the detector corresponding to position 4 in Figure 4 can be computed in accordance with Formula (1).

$$P_{00} = \sum_{i=1}^{n} (P_i \times SR_i)/SR_{bd}, \quad (1)$$

The detector output power P_{01} at position 5 of Figure 4 can be calculated by Formula (2).

$$P_{01} = \sum_{i=1}^{n} (P_i \times SR_m)/SR_m = \sum_{i=1}^{n} P_i, \quad (2)$$

Divide P_{00} and P_{01} to obtain the ratio K_P of the output power of the single-channel detector before and after eliminating the spectral responsivity.

$$K_P = \sum_{i=1}^{n} (P_i \times SR_i)/(SR_{bd} \times \sum_{i=1}^{n} P_i), \tag{3}$$

where SR and SR_{bd} are fixed constants for the same type of detector. The modified spectral responsivity SR_m of the detector after the coating window is added is also a constant. $\sum_{i=1}^{n} P_i$ is recorded as the total power P_{total} of the incident laser, and the power P_i (i = 1~n) at each wavelength is recorded as the proportion $K_i \times P_{total}$ relative to the total power. Therefore, Formula (3) can be further written as

$$\begin{aligned} K_P &= P_{total} \times \sum_{i=1}^{n} (K_i \times SR_i)/(SR_{bd} \times P_{total}) \\ &= \frac{\sum_{i=1}^{n} (K_i \times SR_i)}{SR_{bd}} \end{aligned}, \tag{4}$$

Therefore, the ratio K_p of the output power before and after the single-channel detector eliminates the quantum difference is a fixed coefficient, and the output response value of the detector when a coated window mirror is installed at the front end of the detector can be seen as the real laser power at this point. This means that the laser power of each detector channel can be corrected by K_p, so as to solve the problem of expanding the incident laser wavelength range of the detector array target. The calculation method is shown in Formula (5).

$$P(f) = \frac{\sum_{i=1}^{m-1} \sum_{j=1}^{n-1} A_{ij} I_{ij}(f) + 2 \times A_{00} I_{00}(f)}{K_P} \tag{5}$$

where m represents the number of rows of the detector array, n represents the number of columns of the detector array, A_{ij} represents the area (unit is cm^2) represented by the detection unit with rows and columns (i, j), $I_{ij}(f)$ is the light intensity output value of the point, and A_{00} and $I_{00}(f)$ are the area and light intensity output values of the detector at the corresponding position of the coated window mirror.

With regard to the issue of mitigating the spectral response difference of the detector through the application of the coating method, the most crucial influencing factor lies in the design of the film layer. For the detector mentioned above, within the spectral band ranging from 1 μm to 1.1 μm, the coating layer designed by our research group has managed to reduce the spectral response difference from 4% to approximately 1.7%. Nevertheless, there remains substantial room for further optimization in the design of this layer.

4. Conclusions

In summary, as elucidated by the foregoing analysis, the spectral responsivity curve is acquired through the measurement of the detector's quantum efficiency. On this basis, the design of the corresponding film is capable of effectively mitigating the spectral responsivity difference of the detector. Within the measurement system, the incident light in a single channel is bifurcated into two portions; one is irradiated onto the detector at its original corresponding position, and for the other the detector is furnished with a coated window mirror. The correction coefficient can be derived by dividing the two sets of data, thereby facilitating the precise measurement of the laser intensity distribution within the measurement system. This methodology not only substantially curtails the research and development expenditure of the measurement system while addressing the spectral response difference of the detector but also optimizes the engineering complexity associated with system implementation and affords significant convenience to data processing operations. Notably, this method is applicable to large-scale detection arrays and is not circumscribed to the domain of laser parameter measurement.

Author Contributions: Conceptualization, Y.C. and G.W.; methodology, F.T. and Z.H.; software, Y.C., F.H. and Y.L.; writing—original draft preparation, Y.C. and G.W.; writing—review and editing, F.T., Z.H. and L.Q. All authors have read and agreed to the published version of the manuscript.

Funding: This research received no external funding.

Institutional Review Board Statement: Not applicable.

Informed Consent Statement: Not applicable.

Data Availability Statement: Data are available upon request.

Conflicts of Interest: The authors declare no conflicts of interest.

References

1. Fedoseev, V.I. Optimizing the signal processing of detector arrays, using the window method. *J. Opt. Technol.* **2010**, *77*, 272–279. [CrossRef]
2. Feng, Y.; Vinogradov, I.; Ge, N.-H. Optimized noise reduction scheme for heterodyne spectroscopy using array detectors. *Opt. Express* **2019**, *27*, 20323–20346. [CrossRef] [PubMed]
3. Guan, W.L.; Tan, F.F. Wide angle array detection technology for high power density laser. *Acta Opt. Sin.* **2022**, *42*, 159–166.
4. Liu, Y.; Zeng, X.; Cao, C.; Feng, Z.; Lai, Z.; Feng, Y.; Vinogradov, I.; Ge, N.-H. Target speckle correction using an array detector in heterodyne detection Optimized noise reduction scheme for heterodyne spectroscopy using array detectors. *Opt. Lett.* **2019**, *44*, 5896–5899. [CrossRef] [PubMed]
5. Prigent, S.; Dutertre, S.; Bidaud-Meynard, A.; Bertolin, G.; Michaux, G.; Kervrann, C. Sparse denoising and adaptive estimation enhances the resolution and contrast of fluorescence emission difference microscopy based on an array detector. *Opt. Lett.* **2023**, *48*, 498–501. [CrossRef] [PubMed]
6. He, X.; Xiang, Y. Study on a method of evaluating the alignment of pixels between fiber-optic image bundles and detector arrays. *Appl. Opt.* **2011**, *50*, E189–E192. [CrossRef]
7. Zhang, Z.; Ye, Z.; Song, D.; Zhang, P.; Chen, Z. Repositioning and steering laser beam power via coherent combination of multiple Airy beams. *Appl. Opt.* **2013**, *52*, 8512–8517. [CrossRef] [PubMed]
8. Zhao, P.; Dong, Z.; Zhang, J.; Lin, X. Passive coherent beam combination of three Nd:YAG lasers using cascaded Michelson-type compound cavities. *Opt. Express* **2018**, *26*, 18019–18027. [CrossRef] [PubMed]
9. Fedoseev, V.N.; Zhupanov, V.G. Design of optical Fabry–Perot filters for spectral combination of laser beams. *J. Opt. Technol.* **2021**, *88*, 683–687. [CrossRef]
10. Tao, R.; Si, L.; Ma, Y.; Zhou, P.; Liu, Z. Coherent beam combination of fiber lasers with a strongly confined waveguide: Numerical model. *Appl. Opt.* **2012**, *51*, 5826–5833. [CrossRef] [PubMed]
11. Park, S.; Cha, S.; Oh, J.; Lee, H.; Ahn, H.; Churn, K.S.; Kong, H.J. Coherent beam combination using self-phase locked stimulated Brillouin scattering phase conjugate mirrors with a rotating wedge for high power laser generation. *Opt. Express* **2016**, *24*, 8641–8646. [CrossRef] [PubMed]
12. Chang, W.Z.; Zhou, T.; Siiman, L.A.; Galvanauskas, A. Femtosecond pulse spectral synthesis in coherently-spectrally combined multi-channel fiber chirped pulse amplifiers. *Opt. Express* **2013**, *21*, 3897–3910. [CrossRef] [PubMed]
13. Shi, M.; Yu, M.; Fang, Z.; Wu, Y.; Li, J.; Wang, J.; Mu, H.; Hu, W.; Yi, L. Real-Time Definite Sequence Modulation Based Spectral Broadening Scheme for High-Power Narrow-Linewidth Fiber Laser. *J. Light. Technol.* **2022**, *40*, 6222–6229. [CrossRef]
14. Eppeldauer, G.P.; Podobedov, V.B. Infrared spectral responsivity scale realization and validations. *Appl. Opt.* **2012**, *51*, 6003–6008. [CrossRef] [PubMed]
15. López, M.; Hofer, H.; Stock, K.D.; Bermúdez, J.C.; Schirmacher, A.; Schneck, F.; Kück, S. Spectral reflectance and responsivity of Ge- and InGaAs-photodiodes in the near-infrared: Measurement and model. *Appl. Opt.* **2007**, *46*, 7337–7344. [CrossRef] [PubMed]
16. Yuan, L.; Qiu, L. Wavelength calibration methods in laser wavelength measurement. *Appl. Opt.* **2021**, *60*, 4315–4324. [CrossRef] [PubMed]

17. Yang, X.; Jia, M.; Xiao, G.; Chai, Q.; Zhou, R.; Romashko, R.V.; Zhang, J. FBG array based wavelength calibration scheme for Fourier domain mode-locked laser with pm resolution and hourly stability. *Opt. Express* **2022**, *30*, 45393–45399. [CrossRef] [PubMed]
18. Stair, R.; Schneider, W.E.; Waters, W.R.; Jackson, J.K. Some Factors Affecting the Sensitivity and Spectral Response of Thermoelectric (Radiometric) Detectors. *Appl. Opt.* **1965**, *4*, 703–710. [CrossRef]
19. Yao, P.; Tu, B.; Xu, S.; Yu, X.; Xu, Z.; Luo, D.; Hong, J. Non-uniformity calibration method of space-borne area CCD for directional polarimetric camera. *Opt. Express* **2021**, *29*, 3309–3326. [CrossRef] [PubMed]

photonics

MDPI

Article

Research on an Echo-Signal-Detection Algorithm for Weak and Small Targets Based on GM-APD Remote Active Single-Photon Technology

Shengwen Yin [1], Sining Li [1], Xin Zhou [1,2,3,*], Jianfeng Sun [1,2], Dongfang Guo [1], Jie Lu [4] and Hong Zhao [1]

[1] National Key Laboratory of Laser Spatial Information, Harbin Institute of Technology, Harbin 150001, China
[2] Zhengzhou Research Institute, Harbin Institute of Technology, Zhengzhou 450000, China
[3] Research Center for Space Optical Engineering, Harbin Institute of Technology, Harbin 150001, China
[4] 44th Research Institute of China Electronics Technology Group Corporation, Chongqing 400060, China
* Correspondence: zx2021@hit.edu.cn

Abstract: Geiger-mode avalanche photodiode (GM-APD) is a single-photon-detection device characterized by high sensitivity and fast response, which enables it to detect echo signals of distant targets effectively. Given that weak and small targets possess relatively small volumes and occupy only a small number of pixels, relying solely on neighborhood information for target reconstruction proves to be difficult. Furthermore, during long-distance detection, the optical reflection cross-section is small, making signal photons highly susceptible to being submerged by noise. In this paper, a noise fitting and removal algorithm (NFRA) is proposed. This algorithm can detect the position of the echo signal from the photon statistical histogram submerged by noise and facilitate the reconstruction of weak and small targets. To evaluate the NFRA method, this paper establishes an optical detection system for remotely detecting active single-photon weak and small targets based on GM-APD. Taking unmanned aerial vehicles (UAVs) as weak and small targets for detection, this paper compares the target reconstruction effects of the peak-value method and the neighborhood method. It is thereby verified that under the conditions of a 7 km distance and a signal-to-background ratio (SBR) of 0.0044, the NFRA method can effectively detect the weak echo signal of the UAV.

Keywords: GM-APD; weak and small target detection; signal-detection algorithm

1. Introduction

Weak and small targets are prominently characterized by weakness and smallness. The International Society for Optics and Photonics has previously defined a target with a pixel proportion not exceeding 0.12% as a small target [1]. When weak and small targets are imaged over long distances, their echo signals are submerged by noise after atmospheric attenuation, resulting in a low signal–noise ratio of the data. Additionally, interference sources in nature can readily trigger false alarms. Weak and small targets have small imaging cross-sectional areas. During the detection process, they only possess weak information regarding shape, size, and structure and simultaneously lack texture content, making it difficult to establish an exact segmentation boundary between the target and the background. These reasons result in the inapplicability of traditional signal-detection techniques to weak and small targets, thereby making it difficult to reconstruct such targets.

In recent years, the detection of echo signals of weak and small targets has garnered significant academic attention [2,3]. Nevertheless, numerous challenges in the detection of weak and small targets arise from the limitations of detector systems, equipment size, weight, operation modes, and detection principles [4]. The use of traditional methods, such as infrared and visible light, for detecting weak and small targets is highly susceptible to environmental factors [5], accompanied by a limited detection range and substantial difficulty in detecting in complex backgrounds [6], thereby making it challenging to separate the target from the background. The single-photon imaging-detection technology based on GM-APD features single-photon-level detection sensitivity and nanosecond-level time resolution [7–9]. By adopting an active illumination mode [10,11], this technology can conduct high-precision three-dimensional imaging of the target scene [12–14], thus fulfilling the detection requirements of long-distance, nighttime, and complex backgrounds. The active-type transmitting laser can shield against electromagnetic interference [15,16], possess all-weather detection capabilities, and constitutes an ideal solution for realizing long-distance optical imaging.

In the context of signal detection from weak and small targets, detecting the real-distance information from the TOF (Time-of-Flight) statistical histogram obtained by the single-photon lidar-detection system poses a significant challenge. When only a limited number of signal photons reach the detector, environmental photons and noise can overwhelm the signals. To address this challenge, numerous heuristic algorithms have been proposed for handling the detection of weak target signals. In article [17], signal detection is achieved by setting a short-duration sliding window and seeking the largest signal cluster among the signal clusters. This is because signal detection is more prone to forming clusters than background detection, and the addition of a sliding window can enhance the prominence of the signal clusters. For pixels with insufficient detection times to accurately execute the short-duration range gate, data from adjacent pixels are combined to enhance depth estimation. Article [18] has developed an array-specific algorithm. By leveraging the lateral smoothness and longitudinal sparsity of the natural scene, it transforms the coarse-time-bin photon detection into high-precision scene depth and reflectivity. By surmounting the coarse-time resolution limitation of the array, the framework attains high photon efficiency within a relatively short acquisition time. Study [19] has proposed a deep-learning signal-detection scheme. A convolutional neural network (CNN) is developed and trained, using the array photon-signal-count histogram as the data input. A three-dimensional convolution kernel is employed to process the input data at multiple resolution scales, and subsequently, the outputs are fused to generate the results of photon-signal detection. Nevertheless, these algorithms overly rely on neighborhood information and are unable to properly handle the excessive noise in the front position of the gating gate, which fails to align with the single-photon data characteristics of weak and small targets. Consequently, their performance in signal detection from weak targets is less than satisfactory.

With the increase in detection distance, the signal photons returning from the target decline sharply. Additionally, atmospheric disturbances, scattering, sunlight noise, and detector dark current all introduce noise [20], thereby leading to an extremely low signal-to-noise ratio of the echo signal. Moreover, the proportion of target pixels within the entire field of view is relatively small, which makes it challenging to utilize neighborhood information. There remain substantial challenges in achieving the detection of signals from weak and small target with single-photon imaging technology. The NFRA method is proposed for detecting long-distance signals from weak and small target. The law of background noise and the temporal sequence characteristics of weak signals are utilized to extract and reconstruct the weak signals of weak and small targets, and the effectiveness of the NFRA method in detecting signals from weak and small target is analyzed. Finally, a long-distance active single-photon signal-detection system for weak and small targets based on the GM-APD single-photon detector is constructed to verify the feasibility of this scheme.

2. Proposed Algorithm

2.1. The Trigger Probability Model of GM-APD Detector

GM-APD detectors possess extremely high photon-response capabilities. The laser emits a periodic light beam directed towards the target. After the photons irradiate the target surface, they return and are then received by the detector. At the single-photon level, the avalanche effect of the GM-APD detector can be induced [21,22]. Nevertheless, this also leads to the inability of the GM-APD detector to distinguish whether the avalanche current is triggered by echo photons or noise photons.

During each laser pulse period, upon being triggered by the first received photon, the GM-APD detector generates an avalanche current, thereby triggering photon counting. During a single laser pulse, the detector can be triggered at most once and then enters the dead time. When waiting for the start of the next laser period, the detector resumes its operation. In fact, a single laser pulse obtains a time of flight.

In operation, the GM-APD often adopts the method of multi-frame accumulation, and multiple sets of time of flight are statistically compiled into a histogram. The abscissa of the histogram represents the photon time of flight, which is discretized into a series of equally wide intervals at certain time intervals (bin), and the ordinate represents the multiple statistical results corresponding to the same time of flight, that is, the pulse amplitude. Within the gate width, the time is partitioned into bins of the same size, and the time when photon counting occurs is recorded by a time-stamp. $P(\text{n}; \text{i})$ can be utilized to represent that the avalanche effect is triggered n times within the i-th bin. The echo waveform cannot be acquired through a single time detection and requires continuous acquisition for obtaining.

In the laser-detection scenario of this study, although at the level of a single laser emission it seems to be a binomial situation where a pulse is either detected or not detected, in reality, photons will be affected by multiple factors such as atmospheric scattering and the reflection characteristics of the target during the propagation process, resulting in the randomness of the process in which photons reach the detector. For example, even under the same experimental conditions, there will be certain fluctuations in the number of photons returning and their arrival times after each laser emission. The Poisson distribution is applicable to describing the number of occurrences of random events within a certain time or space, and these events occur independently of each other. In laser detection, the event of each photon arriving at the detector can be regarded as relatively independent, and in the long-term statistical process, the overall situation of photon arrivals conforms more to the characteristics of the Poisson distribution. The probability of n-time triggering within the i-th bin is as follows [23]:

$$P(n;\text{i}) = \frac{1}{n!}[M(\text{i})]^n \exp[-M(\text{i})] \tag{1}$$

During a single laser pulse period, the photon count is triggered in the f-th bin, while the preceding $f-1$ bins have no photon count triggered. Therefore, the triggering probability can be expressed as follows [24,25]:

$$\begin{aligned} P_f &= \left[\prod_{i=1}^{f-1} P(n=0;i)\right] P(n \geq 1; f) \\ &= \exp\left(-\sum_{i=1}^{f-1} M(i)\right)[1-\exp(-M(f)] \end{aligned} \tag{2}$$

Let N denote the average noise within each bin interval and S represent the signal photons within the gate width. Consequently, the probability of triggering the photon count within the target bin can be rewritten as follows:

$$P_f = \exp(-fN)[1-\exp(-S-N)] \tag{3}$$

The probability of detecting a signal in bin f is equal to the probability of not detecting a signal in the previous $f-1$ bins, multiplied by the probability of detecting a signal in the f-th bin. In other words, the triggering probability of echo photons is related not only to the current interval but also to the state that there is no photon-counting trigger in all the previous intervals.

2.2. Signal-Detection Algorithm

The laser emission system irradiates the target surface with a laser of a fixed period interval. Photons returning from the target are incident on the GM-APD detector, thereby triggering an avalanche current for the generation of photon counting. Photon counting enables the recording of the arrival time of each photon, and after long-term statistical analysis, a histogram statistical result can be generated. The reflection intensity and depth at a specific pixel can be analyzed based on the amplitude of the observed pulse signal and the average time of flight.

Owing to the small reflection cross-section of the weak and small target and the low signal-to-noise ratio of the signal echo, the number of pixels occupied on the detector is remarkably small. When the light incident on the detector is rather weak, traditional photoelectric detectors with low sensitivity fail to detect the signal successfully. In this regard, a high-sensitivity GM-APD detector is requisite for resolving the problem of single-photon detection, followed by continuous accumulation. Subsequently, the relationship between photon counting and bin is presented in the form of a photon-counting histogram.

The GM-APD single-photon lidar operates in the range-gate mode with a gate width of 2 μs, which corresponds to a distance of 300 m and can cover the distance range of most targets. The detection counts of all pixels are summarized to generate the total raw data histogram h(x), as shown in Figure 1a. Meanwhile, a fitting curve, y_{fine}, is established (see Equation (4)).

$$y_{fine} = a_n x^n + a_{n-1} x^{n-1} + \ldots + a_1 x + a_0 \tag{4}$$

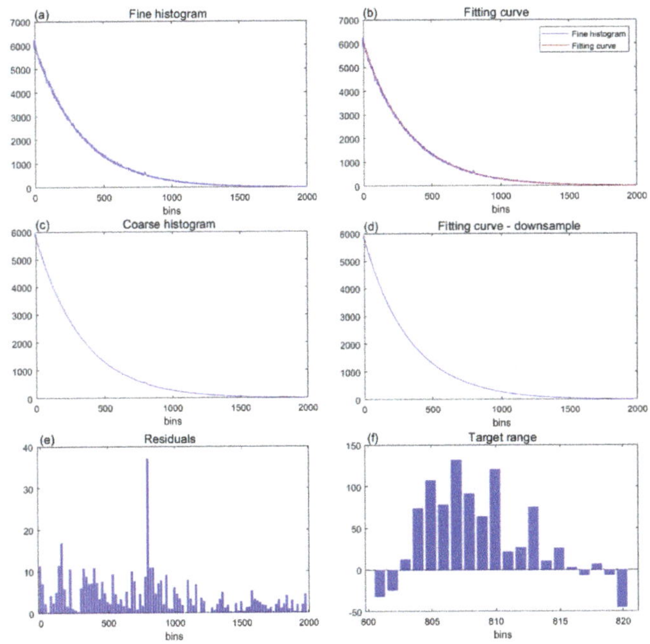

Figure 1. (**a**) Fine histogram of raw data; (**b**) fitting curve based on fine histogram; (**c**) down-sampling the fine histogram; (**d**) down-sampling the fitting curve; (**e**) residuals histogram; (**f**) target signal position interval.

This is an n-order polynomial, where a_n, a_{n-1}, . . ., a_0 are the coefficients of the polynomial, and n is the highest degree of the polynomial. Then, polynomial fitting of the fitting curve is performed according to the least-squares method (see Equation (5)).

$$min_{a_n, a_{n-1}, \ldots, a_0} \sum_{i=1}^{m} \left(h(x_i) - y_{fine}(x_i) \right)^2 \tag{5}$$

Among them, the x_i corresponds to the sampling point of the abscissa in the histogram data. The finally obtained fitting curve is shown in Figure 1b. The relative standard deviation (RSD) between the original data and the fitting curve is 2.58%. Background noise encompasses ambient light and dark counts. The noise of the lidar near the starting position of the gate width is greater since the triggering probability of noise photons at the front end of the gate width is higher. The original data histogram h(x) and the fitting curve y_{fine} are down-sampled, as shown in Equation (6). The step size is set as 20 ns (~3 m) for generating a coarse histogram and a coarse fitting curve, as depicted in Figure 1c,d.

$$h_{coarse}(x_i) = \sum_{D*(i-1)+1}^{D*i} h(x_j)/D; \; y_{coarse}(x_i) = \sum_{D*(i-1)+1}^{D*i} y_{fine}(x_j)/D \tag{6}$$

Among them, x represents the integer index of the abscissa, and D represents the down-sampling factor. By subtracting the coarse histogram h_{coarse} from the coarse fitting curve y_{coarse}, a residual curve R is obtained, as shown in Figure 1e. From this, the position index corresponding to the maximum residual is located (see Equation (7)).

$$x_k = \text{argmax}\{R(x_1), R(x_2), \ldots, R(x_n)\} \tag{7}$$

The coordinate index x_k corresponding to the maximum residual is expanded within the range of the down-sampling step D in the original histogram h(x) to obtain the target signal position interval, namely the target range. Due to the random nature of noise, there may be instances where the statistical amplitude is higher than that of the target echo. Nevertheless, noise lacks regularity in the time domain. Within the target range, the matched filtering method can exploit the regularity of the statistical results of the local region of the signal, thereby increasing the amplitude of the echo signal and enhancing the accuracy of target-signal detection.

3. Experiment

3.1. System Design

Figure 2 presents a long-range active single-photon UAV-detection system based on GM-APD. In this system, the GM-APD detector functions as the receiving end, while a laser acts as the transmitting end. The optical signal irradiates the target surface from the transmitting end via the transmitting lens, and the reflected signal is received by the GM-APD detector through the reflecting lens and the optical system. The laser transmitting unit commences with the laser, and its operating wavelength is selected as 1064 nm. The transmitting beam formed by the beam expander irradiates the entire field of view, thereby enhancing the detection efficiency of the 64 × 64-pixel APD array. A high-repetition-frequency laser is utilized to augment the accumulation of echo signals within a unit time. The echo signal of the lidar is proportional to the transmitted laser power (pulse energy × repetition rate). In other words, reducing the single-pulse energy and increasing the repetition rate can achieve the same detection effect. The high repetition rate not only facilitates the accumulation of echo signals within a unit time, but also significantly improves the signal-to-noise ratio at low-pulse energy [26].

The receiving optical path is designed using a Cassegrain-type telescope system. The folding-back design can reduce the volume resulting from the long optical path. A long focal length is advantageous for detecting distant targets, and a small field of view facilitates distinguishing small distant targets and suppressing external optical noise. The optical components are assembled on a custom-made optical flat plate, and the entire lidar imaging system is integrated to facilitate easy movement.

Overall, the entire detection system features coaxial transmitting and receiving beams, and the transmitting field of view is slightly larger than the receiving one to ensure that the laser spot fills the entire field of view. The receiving field of view is designed to be small, enabling it to effectively filter out external stray light and distinguish unmanned aerial vehicle targets at a long distance. The returning photons are coupled to the focal plane of the GM-APD detector via the focusing lens. During the actual target-detection process, a range gate is set to further filter out irrelevant noise.

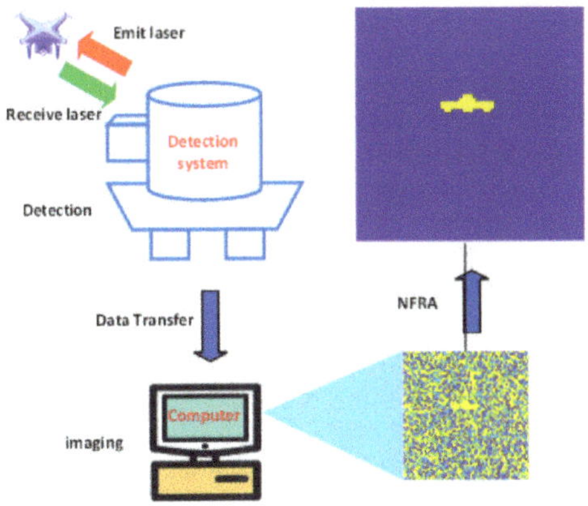

Figure 2. Schematic diagram of long-range active single-photon UAV detection.

3.2. Experiment

In this paper, in-depth research on the imaging system and signal-detection algorithm has been conducted to adapt to the detection of UAV targets within a long-distance space range. The UAV is launched from the equipment, and the distance between the UAV and the equipment is determined by both the linear distance and the height distance between them. The NFRA method was employed to reconstruct the range image of the target. The SBR is defined as the ratio of the signal-detection count (the number of echo photons of the target) within the 2000 ns gate width in the original data to the noise-detection count (the number of ambient light photons and dark counts).

As shown in Figure 3a, this is the experimental environment for UAV detection. The imaging device is positioned on the top floor of a building, while the target is in the distant sky. The experimental targets are the DJI Air 3 (Figure 3b) and the DJI Phantom 4 (Figure 3c).

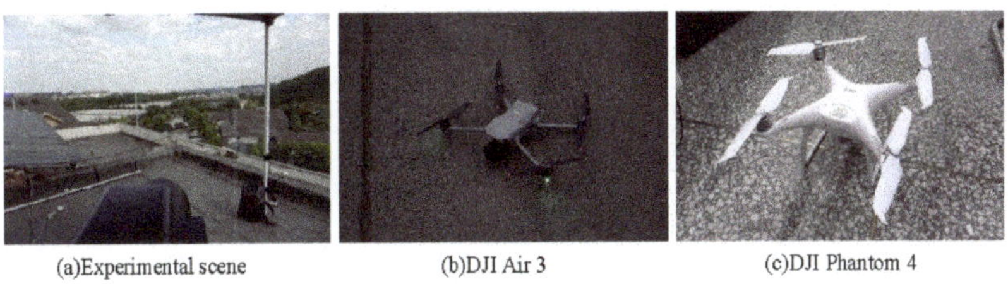

(a)Experimental scene (b)DJI Air 3 (c)DJI Phantom 4

Figure 3. (**a**) Experimental scene, (**b**) the DJI Air 3 UAV with a length of 258.8 mm, a width of 326 mm, and a height of 105.8 mm; (**c**) the DJI Phantom 4 UAV with a length of 430 mm, a width of 430 mm, and a height of 370 mm.

4. Results

Due to insufficient spatial resolution and the influence of air turbulence (aerosol), UAV is undetectable by the human eye and visible-light cameras. Single-photon lidar was utilized for imaging during the daytime, obtaining 64 × 64-pixel images. First, the DJI Air 3 was selected for detection and imaging at a distance of 1 km. The experimental results are presented in Figure 4, and the target echo SBR is 0.035. The original data histogram is shown in Figure 4a. The noise peak near the starting position of the range gate is significantly high and exhibits a continuous attenuation state. By observing the statistical results of the histogram of a single pixel (as shown in Figure 4b), it can be seen that the signal is submerged in the noise, and thus the signal cannot be directly detected by the peak-value method for image reconstruction. A noise threshold needs to be set for processing purposes, as shown in Figure 4c,d. The reconstruction results indicate that the contour of the target cannot be restored, and a large amount of noise cannot be filtered out. Due to its detection capability at the single-photon level, a single-photon detector will generate crosstalk noise during the detection process, and photons will trigger adjacent detector pixels. The neighborhood method increases the detection probability of echo signals by jointly counting the values of neighboring pixels. Through the neighborhood method, most of the noise can be filtered out and part of the contour can be restored as well, as shown in Figure 4e,f. However, there are still instances where the signal detection of pixel points is inaccurate, and the contour information cannot be completely restored. In the time domain, signal photons have a relatively high triggering probability in the target area, while noise photons are discretely distributed due to their randomness. Therefore, the position of the signal peak can be highlighted by the means of convolution in the time domain. Through the combination of the neighborhood method and the convolution method, a good contour restoration effect is presented, yet there are still some instances where the noise cannot be completely filtered out, as shown in Figure 4g,h. By employing the NFRA method, the noise can be effectively suppressed, the real area of the echo signal can be detected, the overall shape of the UAV can be restored, and an excellent reconstruction effect is achieved, as shown in Figure 4i,j.

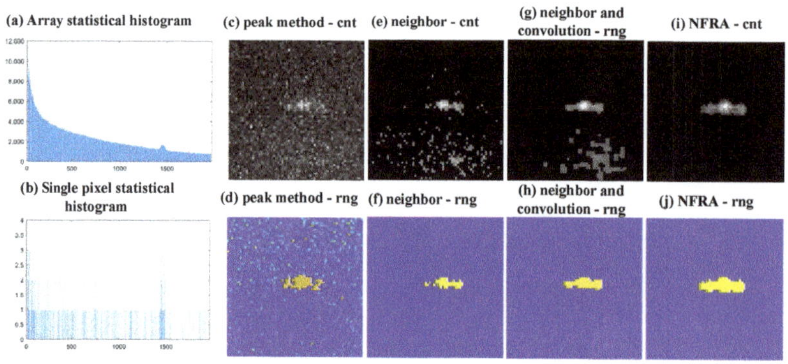

Figure 4. (**a**) Array statistical histogram; (**b**) single-pixel statistical histogram; (**c**–**j**) reconstruction results of the DJI Air 3 UAV at a distance of 1 km using different methods.

To verify the detection effect of the long-range active single-photon unmanned aerial vehicle (UAV)-detection system on distant UAVs, the DJI Phantom 4 was adopted as the target for detection experiments at a distance of 7 km. The reconstruction results are presented in Figure 5. Since the imaging field-of-view angle of the long-range active single-photon UAV-detection system is fixed, when the UAV is at a long distance, the resolvable size of a single pixel within the detection system increases. Moreover, the UAV itself is relatively small in size, resulting in the actual imaging echo signal of the UAV being concentrated only in a few pixels. The surface of the UAV causes diffuse reflection of the irradiated laser, and aerosols impede the transmission of the echo. Furthermore, a large number of noise photons are received by the GM-APD detector array, and the SBR is merely 0.0044. The echo signal is submerged in the noise, making it difficult to directly distinguish the signal from the noise, as shown in Figure 5a,b. For the UAV target at a distance of 7 km, due to the factor of the field-of-view angle of the single-photon-detection system, the number of imaging pixels is extremely small, and the echo signal is extremely weak. With a large amount of noise, it is impossible to directly distinguish the signal from the noise. The processing results of the peak method, the neighbor method, and the combination of the neighbor and convolution methods are accompanied by a large amount of noise, and the target signal cannot be detected. By employing the NFRA method, the noise can be effectively suppressed, and the weak signal of the target can be detected under the condition of being submerged by a large amount of noise, thereby enabling the reconstruction of the UAV target.

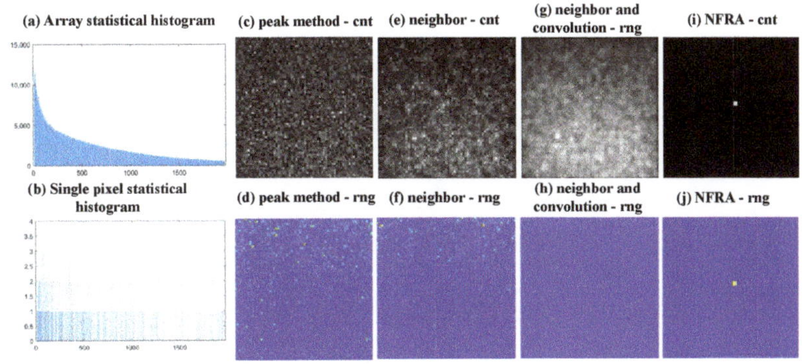

Figure 5. (**a**) Array statistical histogram; (**b**) single-pixel statistical histogram; (**c**–**j**) reconstruction results of the DJI Phantom 4 UAV at a distance of 7 km using different methods.

The NFRA method exhibits better extraction and reconstruction capabilities for the echo signals of weak and small targets compared to the peak method and the neighborhood method. The peak method achieves better effects when processing data with a relatively high signal-to-noise ratio. In the face of a large amount of noise with high amplitude, it is easily interfered with by noise and fails to correctly distinguish between signals and noise. Since weak and small targets themselves have an extremely low signal-to-noise ratio and the echo signal is submerged in the noise, it is very difficult for the peak method to detect the echo signal. The neighborhood method depends on capturing the pattern of the target in the spatial neighborhood to filter out most of the noise, thereby reconstructing the target. However, due to the small number of pixels occupied by the weak and small target, there is scant neighborhood information to rely on. Simultaneously, in the face of the interference of random high-amplitude noise, the complete structure of the target cannot be reconstructed. The NFRA method proposed in this paper can separate the background noise from the echo signal region by fitting and removing the noise in the time domain. According to the regular distribution of the echo signal in the time domain, convolution strengthening is performed to suppress the interference of the noise peak, and the shape of the target can be well restored, thereby achieving the purpose of detecting the signal from weak and small targets.

The detection of long-distance, weak and small UAV targets analyzed in this study is based on cooperative targets. Cooperative targets are more advantageous for achieving effective search, detection, and target confirmation in long-range scenarios. If non-cooperative targets intrude during the detection process, false positives may occur. In such cases, the airspace will exhibit non-stationary states characterized by irregular motion patterns, which are relatively easy to distinguish. However, backscattering caused by clouds introduces additional challenges by increasing noise levels and interfering with the detection range. In scenarios with dense cloud cover, there is a risk of detector saturation. Unlike non-cooperative targets, clouds exhibit relatively stable spatial positions within the airspace, making it difficult to distinguish them based solely on motion characteristics. Nevertheless, clouds possess a certain degree of transparency, and their echo signatures differ from those of UAV targets. By accumulating detection results over time and analyzing the distinctive characteristics of echo signals, false positive targets caused by clouds can still be effectively identified and mitigated.

5. Conclusions

GM-APD is a single-photon-detection device characterized by high sensitivity and fast response. It can continuously count the returning photons by relying on ToF to obtain the target echo statistical histogram. In this paper, a noise fitting and removal algorithm (NFRA method) is proposed, which is capable of detecting and reconstructing the weak signals of UAVs. To evaluate the NFRA method, the influence of different distances on the detection of signals from weak and small targets is analyzed through a comparison of the target extraction effects of the peak method, the neighborhood method, the combination of the neighborhood method and the convolution method, and the NFRA method. Experimental results demonstrate that for the 1 km UAV target, the peak-value method, the neighborhood method, the combination of the neighborhood method and the convolution method, and the NFRA method can all achieve target reconstruction. Among them, the NFRA method can restore the target contour and filter out the background noise based on the detected echo signal. For the 7 km UAV target, due to noise interference, the peak method, the neighborhood method, and the combination of the neighborhood method and the convolution method cannot detect the echo signal. By employing the NFRA method to process the received signal, the target echo signal submerged in the noise can be detected, and the background noise can be filtered to achieve the reconstruction of weak and small targets.

Author Contributions: Conceptualization, S.Y. and S.L.; methodology, J.S. and X.Z.; software, D.G. and J.L.; validation, H.Z.; formal analysis, D.G.; investigation, S.Y. and S.L.; resources, J.S.; data curation, X.Z.; writing—original draft preparation, S.Y.; writing—review and editing, S.Y. and X.Z.; visualization, H.Z. and S.L.; supervision, J.S. and X.Z.; project administration, J.S. and J.L.; funding acquisition, J.S. All authors have read and agreed to the published version of the manuscript.

Funding: This research received no external funding.

Institutional Review Board Statement: Not applicable.

Informed Consent Statement: Not applicable.

Data Availability Statement: The original contributions presented in the study are included in the article, further inquiries can be directed to the corresponding author.

Conflicts of Interest: The authors declare no conflicts of interest.

References

1. Kumar, N.; Singh, P. Small and Dim Target Detection in IR Imagery: A Review. *arXiv* **2023**, arXiv:2311.16346.
2. Zamri FN, M.; Gunawan, T.S.; Yusoff, S.H.; Alzahrani, A.A.; Bramantoro, A.; Kartiwi, M. *Enhanced Small Drone Detection using Optimized YOLOv8 with Attention Mechanisms*; IEEE Access: Piscataway, NJ, USA, 2024.
3. Kashi, R.N.; Prashanth, A.; Kashi, S.R.; Prabhakara, G. A survey and analysis of drone detection systems using a systems approach superposed on scenarios. *Syst. Eng.* **2024**, *27*, 598–636. [CrossRef]
4. Al-lQubaydhi, N.; Alenezi, A.; Alanazi, T.; Senyor, A.; Alanezi, N.; Alotaibi, B.; Alotaibi, M.; Razaque, A.; Hariri, S. Deep learning for unmanned aerial vehicles detection: A review. *Comput. Sci. Rev.* **2024**, *51*, 100614. [CrossRef]
5. Rouhi, A.; Arezoomandan, S.; Kapoor, R.; Klohoker, J.; Patal, S.; Shah, P.; Umare, H.; Han, D. An Overview of Deep Learning in UAV Perception. In Proceedings of the 2024 IEEE International Conference on Consumer Electronics (ICCE), Las Vegas, NV, USA, 6–8 January 2024.
6. Kassab, M.; Zitar, R.A.; Barbaresco, F.; Seghrouchni, A.E.F. Drone Detection with Improved Precision in Traditional Machine Learning and Less Complexity in Single Shot Detectors. *IEEE Trans. Aerosp. Electron. Syst.* **2024**, *60*, 3847–3859. [CrossRef]
7. Wang, M.; Sun, J.; Li, S.; Lu, W.; Zhou, X.; Zhang, H. A photon-number-based systematic algorithm for range image recovery of GM-APD lidar under few-frames detection. *Infrared Phys. Technol.* **2022**, *125*, 104267. [CrossRef]
8. Zhang, Y.; Li, S.; Sun, J.; Liu, D.; Zhang, X.; Yang, X.; Zhou, X. Dual-parameter estimation algorithm for Gm-APD Lidar depth imaging through smoke. *Measurement* **2022**, *196*, 111269. [CrossRef]
9. Ma, L.; Sun, J.; Jiang, P.; Liu, D.; Zhou, X.; Wang, Q. Signal extraction algorithm of Gm-APD lidar with low SNR return. *Optik* **2020**, *206*, 164340. [CrossRef]

10. Ni, H.; Sun, J.; Ma, L.; Liu, D.; Zhang, H.; Zhou, S. Research on 3D image reconstruction of sparse power lines by array GM-APD lidar. *Opt. Laser Technol.* **2024**, *168*, 109987. [CrossRef]
11. Li, M.; Li, Y.; Wang, H. Research on target recognition technology of GISC spectral imaging based on active laser lighting. *Front. Phys.* **2022**, *10*, 999637. [CrossRef]
12. Li, Z.-P.; Ye, J.-T.; Huang, X.; Jiang, P.-Y.; Cao, Y.; Hong, Y.; Yu, C.; Zhang, J.; Zhang, Q.; Peng, C.-Z.; et al. Single-photon imaging over 200 km. *Optica* **2021**, *8*, 344–349. [CrossRef]
13. Zhou, X.; Sun, J.; Fan, Z.; Li, S.; Lu, W. Research on detection performance improvement of polarization GM-APD LiDAR with adaptive adjustment of aperture diameter and spatial correlation method. *Opt. Laser Technol.* **2022**, *155*, 108400. [CrossRef]
14. Li, Z.-P.; Huang, X.; Cao, Y.; Wang, B.; Li, Y.-H.; Jin, W.; Yu, C.; Zhang, J.; Zhang, Q.; Peng, C.-Z.; et al. Single-photon computational 3D imaging at 45 km. *Photonics Res.* **2020**, *8*, 1532–1540. [CrossRef]
15. Xin, C.; Lu, W.; Jiang, P.; Sun, J.; Zhou, X.; Zhang, H.; Wang, Q. Research on GM-APD lidar intense information correction technology based on target distance. In *Applied Optics and Photonics China*; SPIE: Bellingham, WA, USA, 2021; Volume 12065.
16. Wang, M.; Sun, J.; Li, S.; Lu, W.; Zhou, X.; Zhang, H. Research on infrared image guided GM-APD range image recovery algorithm under limited detections. *Opt. Lasers Eng.* **2023**, *166*, 107579. [CrossRef]
17. Rapp, J.; Goyal, V.K. A Few Photons Among Many: Unmixing Signal and Noise for Photon-Efficient Active Imaging. *IEEE Trans. Comput. Imaging* **2016**, *3*, 445–459. [CrossRef]
18. Shin, D.; Xu, F.; Venkatraman, D.; Lussana, R.; Villa, F.; Zappa, F.; Goyal, V.K.; Wong, F.N.C.; Shapiro, J.H. Photon-efficient imaging with a single-photon camera. *Nat. Commun.* **2016**, *7*, 12046. [CrossRef]
19. Lindell, D.B.; O'Toole, M.; Wetzstein, G. Single-Photon 3D Imaging with Deep Sensor Fusion. *ACM Trans. Graph. TOG* **2018**, *37*, 113. [CrossRef]
20. Chen, J.; Xie, C.; Ji, J.; Li, L.; Wang, B.; Xing, K.; Zhao, M. Performance Evaluation and Error Tracing of Rotary Rayleigh Doppler Wind LiDAR. *Photonics* **2024**, *11*, 398. [CrossRef]
21. Zhou, X.; Sun, J.; Jiang, P.; Qiu, C.; Wang, Q. Improvement of detection probability and ranging performance of Gm-APD LiDAR with spatial correlation and adaptive adjustment of the aperture diameter. *Opt. Lasers Eng.* **2021**, *138*, 106452. [CrossRef]
22. Liu, D.; Sun, J.; Gao, S.; Ma, L.; Jiang, P.; Guo, S.; Zhou, X. Single-parameter estimation construction algorithm for Gm-APD ladar imaging through fog. *Opt. Commun.* **2021**, *482*, 126558. [CrossRef]
23. Daniel, G.F. Detection and false-alarm probabilities for laser radars that use Geiger-mode detectors. *Appl. Opt.* **2003**, *42*, 5388–5398.
24. Altmann, Y.; Ren, X.; McCarthy, A.; Buller, G.S.; McLaughlin, S. Lidar Waveform-Based Analysis of Depth Images Constructed Using Sparse Single-Photon Data. *IEEE Trans. Image Process.* **2016**, *25*, 1935–1946. [CrossRef] [PubMed]
25. Halimi, A.; Altmann, Y.; McCarthy, A.; Ren, X.; Tobin, R.; Buller, G.S.; McLaughlin, S. Restoration of intensity and depth images constructed using sparse single-photon data. In Proceedings of the 2016 24th European Signal Processing Conference (EUSIPCO), Budapest, Hungary, 9 August–2 September 2016.
26. Cheng, L.; Xie, C. An integrated off-line echo signal acquisition system implemented in SoC-FPGA for high repetition rate lidar. *Electronics* **2023**, *12*, 2331. [CrossRef]

photonics

MDPI

Communication

Multiple Fano Resonances in a Metal–Insulator–Metal Waveguide for Nano-Sensing of Multiple Biological Parameters and Tunable Slow Light

Ruiqi Zhang, He Tian *, Yang Liu and Shihang Cui

College of Science, Northeast Forestry University, Harbin 150040, China; zhangruiqi@nefu.edu.cn (R.Z.); 2020212510@nefu.edu.cn (Y.L.); 458586551@nefu.edu.cn (S.C.)
* Correspondence: tianhe@nefu.edu.cn

Abstract: A surface plasmonic waveguide made of metal–insulator–metal (MIM) capable of generating triple Fano resonances is proposed and numerically investigated for multi-biological parameter sensing as well as tunable slow light. The waveguide is made up of a bus waveguide with a silver baffle, a square split-ring cavity with a square center (SSRCSC), and a circular ring cavity with a square center (CRCSC). Based on the triple Fano resonances, human blood temperature and plasma concentration are measured simultaneously at different locations in the waveguide, and the maximum sensitivities were 0.25 nm/°C and 0.2 nm·L/g, respectively. Furthermore, the two biological parameters can be used to achieve tunable slow light, and it was found that the group delay responses to human blood temperature and plasma concentration all conformed to cubic functions. The MIM waveguide may have great applications in future nano-sensing of multiple biological parameters and information processing of optical chips or bio-optical chips.

Keywords: biosensing; biological parameters; slow light; surface plasmon; MIM waveguide; Fano resonance

1. Introduction

Since Ugo Fano initially proposed it in 1961, Fano resonance has established itself as one of the key components in the advancement of optical sensing [1–4]. Unlike conventional Lorentz resonance with a symmetric lineshape, Fano resonance, which emerges when a broad continuous state is connected to a narrow discrete state, usually has a sharp and asymmetric lineshape. Due to the unique lineshape, small changes in the geometry or environment can significantly affect Fano resonance [5–9]. Electromagnetic waves known as surface plasmon polaritons (SPPs) are found solely between a metal and a nonconductor, at their intersection. Molecular signals can be strengthened by strong SPPs at the metal–dielectric contact site, which is helpful in biomedicine [10–12], chemistry [13,14], and information technology [15]. Because the SPPs also have the excellent characteristic of breaking the classical diffraction limit [16,17], the combination of SPPs and Fano resonance has very important applications in modern micro- and nano-chip optics [18–22].

The SPPs in MIM waveguides, which have adequate propagation lengths, low bend losses, and ease of sample manufacturing, have gained significant study interest [23–26]. Fano-resonance-exciting MIM waveguide designs have gained popularity recently for use in filters [27–29], wavelength division multiplexers [30], all-optical switches [31,32], slow-light devices [33,34], and especially in sensors [6,35–39]. For instance, Chen et al. proposed an SSRC on a MIM waveguide that is non-through and could achieve multiple Fano resonances, and the highest sensing sensitivity for refractive index was up to 1290.2 nm/RIU [38]. A waveguide system with end coupling capable of Fano resonances was designed by Fang et al. The sensing sensitivity of the refractive index could reach 1059.2 nm/RIU [40]. In addition, Wang et al. achieved a square-ring and triangle-cavity

MIM waveguide to obtain triple Fano resonances and a maximal sensitivity to refractive index of 2259.56 nm/RIU [41]. All of these results showed that Fano resonance could be effectively used for integrated optical sensing with high sensitivity.

In this study, we aimed to increase the parallel processing capability of biosensing and obtain tunable slow light, based on easily adjustable multiple Fano resonances. To implement the simultaneous measurement of multiple biological parameters and tuning of slow light effects at selected wavelengths, we proposed multi-ring cavities. This will also make it easier to use Fano resonances in integrated optical biosensing. A MIM surface plasmonic waveguide with ring cavities was designed to generate triple Fano resonances. By altering the refractive index, individual tuning of the triple Fano resonances was examined, and glucose solution concentration and plasma concentration were measured simultaneously using this waveguide. In addition, tunable slow light was achieved using these two biological parameters, and the response of the group delay to these biological parameters was fitted.

2. Materials and Methods

The envisioned waveguide for the MIM waveguide is shown in Figure 1, which is composed of a square split-ring cavity with a square center (SSRCSC), a circular ring cavity with a square center (CRCSC), and a waveguide with a silver baffle for a bus. Silver and air are represented by the white and green regions, respectively. A silver baffle was added to the upper part of the square ring, in order to obtain sufficient distances between Fano resonances for measuring multiple parameters at the same time, and increase the sharpness and transmittance of Fano resonances for realizing larger sensitivity and lower slow light. d, L, a, R, b, G_1 and G_2 are the width of this silver baffle, the side length of the external square of the SSRCSC, the side length of the internal square of the SSRCSC, the radius of the external circle of the CRCSC, the side length of the internal square of the CRCSC, the distance between the SSRCSC and the bus waveguide, and the distance between the CRCSC and the bus waveguide, respectively. To ensure that only the basic transverse magnetic mode may exist in this configuration, the bus waveguide's width w is specified as 50 nm. The silver baffle in the center, whose width is given by t, blocks the bus waveguide. The SSRCSC, CRCSC, and bus waveguide all have geometric centers that are on the reference line, so the overall waveguide is symmetrical about the reference line.

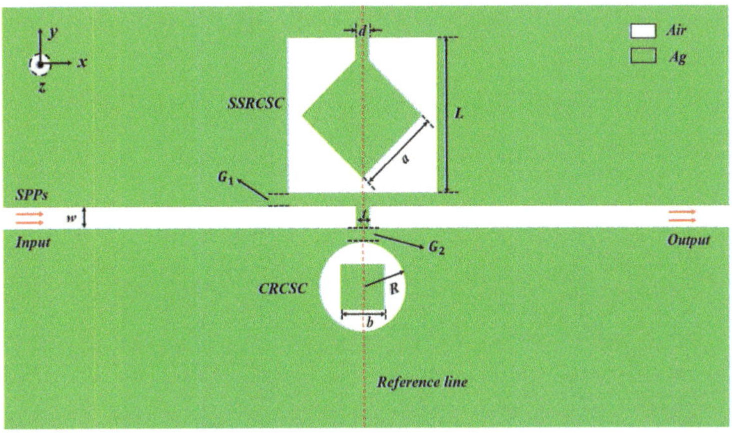

Figure 1. A schematic diagram of a MIM waveguide composed of a square split-ring cavity, a bus waveguide with a silver baffle, and a circular ring cavity.

The Drude model represents the frequency-dependent complicated relative permittivity of silver [42,43]:

$$\varepsilon(\omega) = \varepsilon_\infty - \frac{\omega_p^2}{\omega^2 + i\omega\gamma} \tag{1}$$

where $\varepsilon_\infty = 3.7$, $\omega_p = 9.1$ eV, ω, and $\gamma = 0.018$ eV represent the dielectric constant at infinite frequency, the plasma frequency of free conduction electrons, the angular frequency of the incident wave in vacuum, and the electron collision frequency, respectively.

The standing wave theory states that constructive interference should occur when the resonance condition is satisfied, and the transmitted wavelength is calculated by using the resonance condition [44,45]:

$$\lambda = \frac{2\mathrm{Re}\left(n_{eff}\right) L_{eff}}{m - \frac{\phi}{2\pi}}, m = 1,2,3\ldots, \tag{2}$$

where λ, m, ϕ, $\mathrm{Re}(n_{eff})$, and L_{eff} represent the resonant wavelength, the order of the resonant mode, the phase shift due to reflection, the real part of the effective refractive index, and the effective length of the resonant cavity, respectively.

The waveguide's optical transmission characteristics are simulated by the finite element method (FEM), and the numerical values of the waveguide parameters utilized in the simulation are shown in Table 1. Here, all the parameters were optimized in order to obtain multiple Fano resonances with high transmittance. In order to absorb the escaping waves, perfect matching layers (PMLs) are positioned at the waveguide's top and bottom. Moreover, fine triangular meshes with a maximum size of 10 nm were chosen to provide precise area segmentation in the simulation. In practice, the MIM waveguide may be fabricated in the following way: First, a thick enough Ag layer is prepared by the chemical vapor deposition (CVD) method on a silicon substrate [46]. Then, the SSRCSC, the CRCSC, and the bus waveguide with a silver baffle are etched on the Ag layer through electron beam etching.

Table 1. A list of the simulation settings for the waveguide.

Parameter	Symbol	Quantity	Unit
Length of the side of the external square of the SSRCSC	L	440	nm
Split length of SSRCSC	d	10	nm
Length of the side of the internal square of the SSRCSC	a	300	nm
The separation between the bus waveguide and the SSRCSC	G_1	10	nm
The radius of the external circle of the CRCSC	R	110	nm
Length of the side of the internal square of the CRCSC	b	140	nm
The separation between the bus waveguide and the CRCSC	G_2	10	nm
The size of the bus waveguide	w	50	nm
The size of the bus waveguide's silver baffle	t	10	nm
Index of refractive of bus waveguide	–	1	–
Index of refractive of SSRCSC	n_1	1	–
Index of refractive of CRCSC	n_2	1	–

3. Results

3.1. Mechanism of Fano Resonance and Distribution of Magnetic Fields

As shown in Figure 2, we set the waveguide in SSRCSC with the CRCSC's mode, single bus waveguide mode, and full waveguide mode to illustrate the mechanism of Fano resonance generation. The bus waveguide with the silver baffle removed produced three transmission dips at 1107 nm, 1849 nm, and 2428 nm, which, as represented by the red line, may be thought of as a narrow discrete state. The blue line represents the state produced by the bus waveguide, which is a wide continuous state. The entire waveguide generated three asymmetric and sharp Fano resonances at 1097 nm, 1795 nm, and 2407 nm, known as FR1, FR2, and FR3, as a consequence of interference between the discrete state and continuous state.

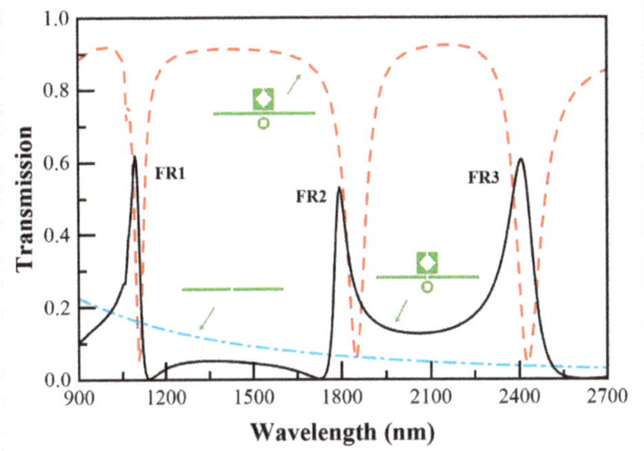

Figure 2. Schematic diagram of the formation mechanism of the structurally excited Fano resonance proposed in this paper.

The dispersion of the magnetic fields ($|H_z|^2$) of FR1, FR2, and FR3 are shown in Figure 3a–c, and the corresponding heights are shown in Figure 3d–f. The distributions of magnetic field energy in the SSRCSC and the CRCSC are symmetric about the reference line parallel to the y-axis. For FR1, almost all of the energy was confined in the CRCSC, so FR1 is sensitive to the parameters of the CRCSC, while the majority of energy of FR2 and FR3 was confined in the SSRCSC, proving that the characteristics of the SSRCSC mostly impact FR2 and FR3. As a result, the SSRCSC and CRCSC characteristics may be individually changed to control the triple Fano resonances.

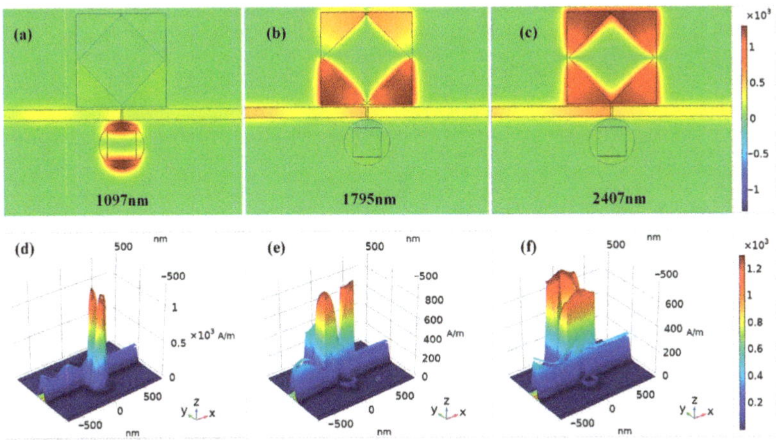

Figure 3. (**a**–**c**) Patterns of the FR1, FR2, and FR3 magnetic fields. (**d**–**f**) Expressions in height for FR1, FR2, and FR3's magnetic field patterns.

3.2. *Refractive Index Sensing*

For practical applications, it is difficult to change the structural parameters of waveguides, but the waveguide's refractive index may be changed much more easily than its structural properties, which makes it possible to accomplish independent tuning of the Fano resonances that are more suited for sensing [47]. The following is the definition of the sensitivity of refractive index sensing [41,48]:

$$S = \frac{\Delta\lambda}{\Delta n}, \quad (3)$$

where $\Delta\lambda$ denotes the alteration in resonance wavelength and Δn denotes the alteration in the index of refraction.

In Figures 4 and 5, the index of refraction n_1 of the SSRCSC and the index of refraction n_2 of the CRCSC both increased from 1.30 to 1.42 with an interval of 0.03. In biological parameter sensing, the chosen range of the index of refraction is easily attained. As the index of refraction of the SSRCSC increases, FR1 remained essentially unchanged, but FR2 and FR3 exhibited significant redshifts. In contrast, only FR1 exhibited a considerable redshift when the CRCSC's refractive index increased, as shown in Figure 5a. According to the findings, it is possible to independently adjust the triple Fano resonances by altering the waveguide's index of refraction.

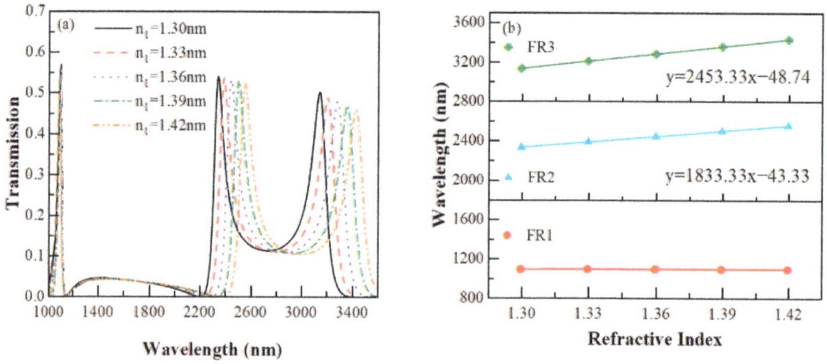

Figure 4. (**a**) The effect of different refractive indexes of SSRCSC on waveguide Fano resonance (n_2 =1.00). (**b**) Associations between the index of refraction of the SSRCSC and the resonance wavelengths of FR1, FR2, and FR3 are linear.

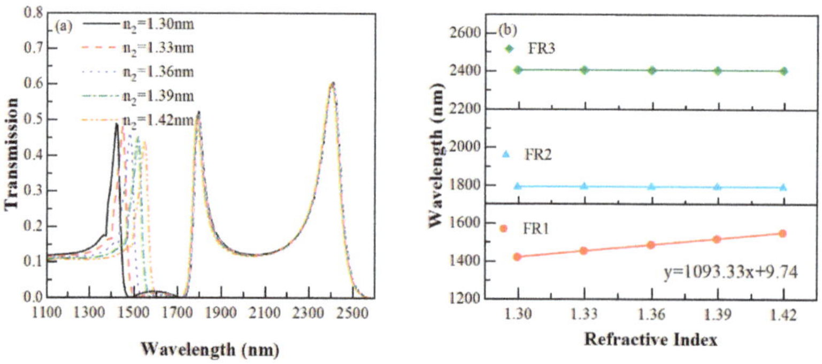

Figure 5. (**a**) The effect of different refractive indexes of CRCSC on waveguide Fano resonance (n_1 = 1.00). (**b**) Associations between the index of refraction of the CRCSC and the resonance wavelengths of FR1, FR2, and FR3 that are linear.

As shown in Figures 4b and 5b, FR1, FR2, and FR3 all possessed strong linear correlations with linear correlation coefficients greater than 0.99999, and the sensitivities of FR1 and FR2 were 1093.33 nm/RIU and 1833.33 nm/RIU, respectively. According to Equations (3) and (4), the sensitivity is proportional to the proportion of the resonance mode order to the effective length of the resonant cavities, so the sensitivity of FR3 was 2453.33 nm/RIU greater than that of FR2, as shown in Figure 4b. Table 2 demonstrates that this waveguide has a relatively high sensitivity to refractive index sensing when compared to other architectures [38,47,49–53]. Based on the above analysis, the refractive index within the waveguide at different positions can be obtained by resonant wavelength measurements.

Table 2. Utilizing different references to compare the sensitivity.

Reference	Waveguide	Sensitivity
[6]	Baffle and an X-shaped cavity make up the MIM waveguide	1303 nm/RIU
[38]	MIM waveguide containing a rectangular split-ring resonance cavity	1290.2 nm/RIU
[47]	MIM waveguide containing a semi-ring cavity	1550.38 nm/RIU
[49]	MIM waveguide containing ring-splitting cavity and tooth cavity coupling	1200 nm/RIU
[50]	MIM waveguide-coupled structure-based simple and small plasmonic sensor	1820 nm/RIU
[51]	Inverted U-shaped resonator	840 nm/RIU
[52]	A MIM waveguide with an end-coupled ring-groove junction	1050 nm/RIU
[53]	Three-racetrack resonators in two concentric rings with plasmonic MIM waveguides	1618 nm/RIU
This paper	MIM waveguide consisting of square split-ring and circular ring cavities	2453.33 nm/RIU

3.3. Multi-Biological Parameter Sensing

Next, two biological parameters, the temperature of human blood and the concentration of plasma, were chosen to be measured using this waveguide. The SSRCSC was filled with human blood, while the CRCSC was filled with plasma. Thus, the refractive indexes of the SSRCSC and the CRCSC were determined by the temperature and concentration of the biological parameters, respectively. In practice, human blood and plasma should be separated in advance, and then filled into the SSRCSC and CRCSC, respectively. In addition, it is unnecessary for the SSRCSC and CRCSC to be fully filled. When they are partially filled with human blood and plasma, the waveguide can still produce similar Fano resonances, just with different resonant wavelengths. The refractive indexes of human blood and plasma are expressed as [54,55]:

$$n_b = 1.36 - 0.0001046T_b \tag{4}$$

$$n_p = 1.32459 + 0.000184C_p \tag{5}$$

where T_b is the temperature of human blood and C_p is the concentration of plasma.

In Figure 6, the temperature of human blood in the SSRCSC was increased from 10 °C to 50 °C with an interval of 10 °C, while the concentration of plasma in the CRCSC was increased from 0 g/L to 400 g/L with an interval of 100 g/L. As a result, the SSRCSC's refractive index dropped from 1.358954 to 1.35477 with an interval of 0.001046, whereas the CRCSC's index of refraction change increased with an interval of 0.0184 from 1.32459 to 1.39819. It is evident that FR1 exhibited a redshift, and FR2 and FR3 exhibited blueshifts.

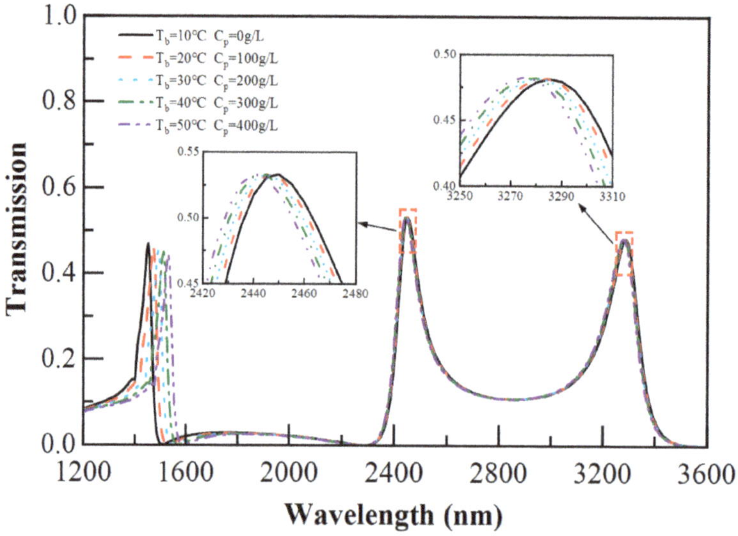

Figure 6. The waveguide's transmission spectra at various plasma concentrations and blood temperatures.

The linear fittings between the concentrations of the plasma and the resonant wavelengths are shown in Figure 7a, and Figure 7b displays the linear association between resonant wavelength and human blood temperature. For FR1, FR2, and FR3, each linear correlation coefficient was greater than 0.99. Here, the sensing sensitivity of the waveguide can be defined as $S_{plasma} = \frac{\Delta\lambda}{\Delta C}$ as well as $S_{human\ blood} = \frac{\Delta\lambda}{\Delta T}$, where ΔC denotes the alteration in concentration and ΔT denotes the alteration in temperature. Thus, 0.2 nm·L/g was the sensitivity of plasma concentration sensing. Moreover, the sensitivity of human blood temperature to FR2 was 0.2 nm/°C, and the sensitivity to FR3 was 0.25 nm/°C. Obviously, the waveguide is far more sensitive to temperature than practically applied fiber grating sensors, which typically have a sensitivity of 0.01 nm/°C. In this way, the concentration of plasma and the temperature of human blood are simultaneously measured using this waveguide.

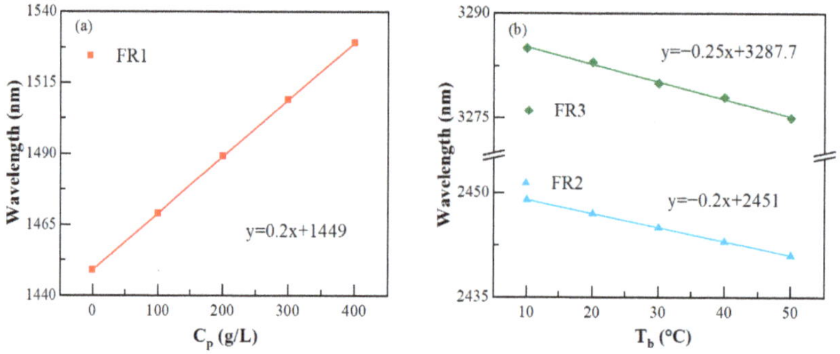

Figure 7. (**a**) FR1's resonant wavelength versus plasma concentration. (**b**) The resonant wavelengths of FR2 and FR3 versus human blood temperature.

3.4. Tunable Slow Light

Due to the sharp and asymmetric lineshape, Fano resonance is accompanied by an abrupt change in transmission phase, resulting in smaller group velocities, so the waveguide designed here can also be used to generate slow light, especially based on the effects of human blood temperature and plasma concentration on Fano resonance, which may achieve tunable slow light. The slow light characteristics can be described in terms of the group delay, which can be derived from the phase:

$$\tau_g = \frac{d\phi(\omega)}{d\omega} \tag{6}$$

As a result of the sharp characteristics of Fano resonance, three group delay peaks can be found near the three Fano resonances, marked as GD1, GD2, and GD3. Then, we verified the proposed conjecture by measuring the biological parameters in the waveguide. In Figure 8, the human blood temperature in the SSRCSC increased with an interval of 10 °C, while the plasma concentration in the CRCSC increased with an interval of 10 g/L. It can be seen that GD1 experienced a redshift, while GD2 and GD3 experienced a blueshift. Subsequently, we locally enlarged GD1, GD2, and GD3, as shown in Figures 9a, 10a, and 11a. We selected the wavelength corresponding to the group delay peak at a temperature of 10 °C and concentration of 0 g/L as the reference wavelength, which was 1508 nm, 2287 nm, and 3585 nm for GD1, GD2, and GD3, respectively. We also plotted the response of the group delay to the biological parameters at each reference wavelength, as shown in Figures 9b, 10b, and 11b. The responses of the group delay to the two biological parameters were not exactly the same. For plasma concentration, it can be seen in Figure 9b that in the initial stage of concentration increase, group delay rapidly decreased. However, after the concentration was high, although the step size of concentration increase was still 10 g/L, the speed of the group delay decreased significantly. In b and 11b, it can be seen that for human blood temperature, the response of GD2 changed slowly during the low and high temperature stages, while the response of GD3 was different. For GD3, at relatively low temperatures, group delay slowly decreased with increasing temperature, and after a certain degree of temperature increase, the response of the group delay to temperature became faster.

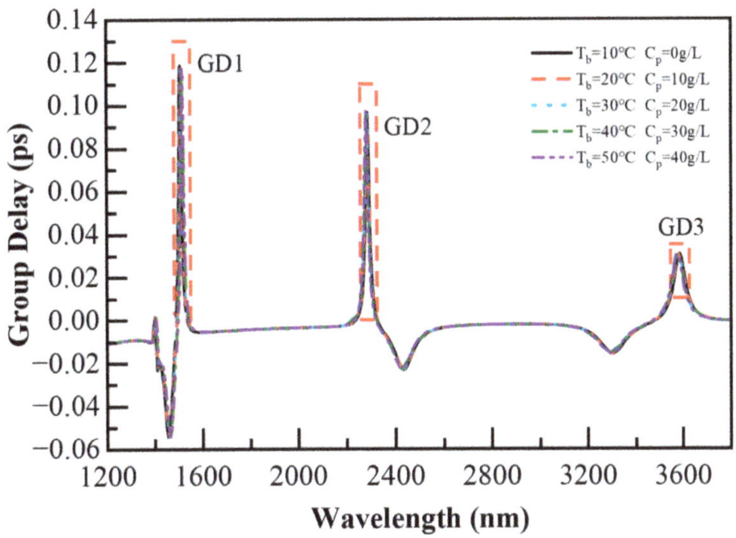

Figure 8. The waveguide's group delay at various plasma concentrations and blood temperatures.

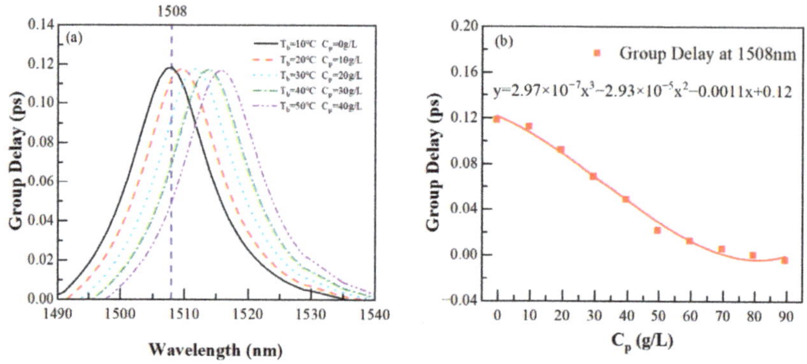

Figure 9. (**a**) Partial enlarged view of GD1. (**b**) Response of group delay at 1508 nm to plasma concentration.

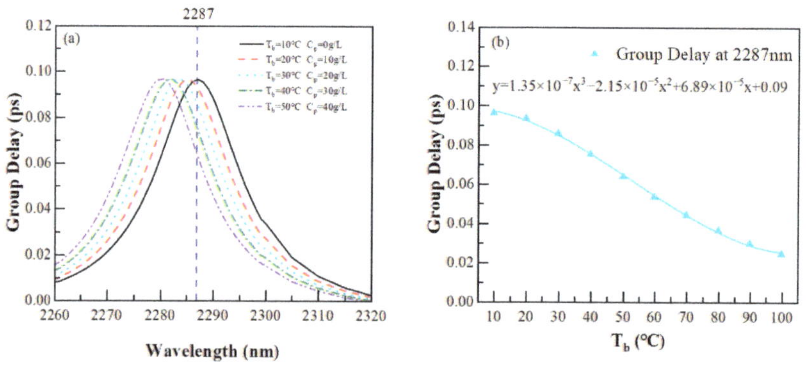

Figure 10. (**a**) Partial enlarged view of GD2. (**b**) Response of group delay at 2287 nm to human blood temperature.

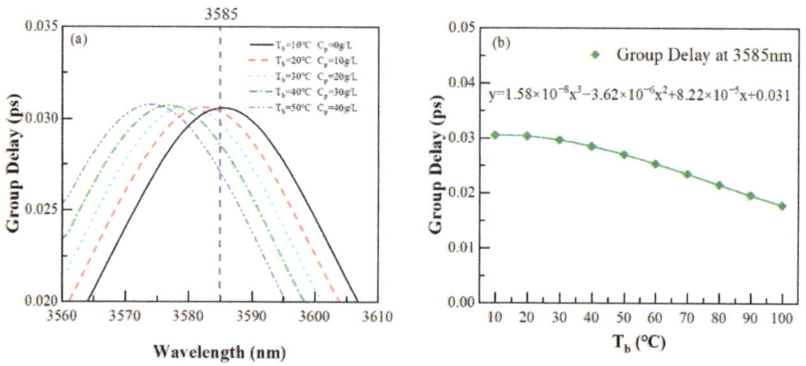

Figure 11. (**a**) Partial enlarged view of GD3. (**b**) Response of group delay at 3585 nm to human blood temperature.

Next, we fit the response of the group delay to these two biological parameters. We first performed a quadratic fit on the response of the group delay to the biological parameters, and the residual sum of squares of the fit results were larger than the expected results. Then, we performed a cubic fit on the three responses, and the residual sum of squares of the fitting curves was much smaller, resulting in a significant improvement in the fitting effect.

The results showed that the two biological parameters can be used to adjust slow light, and it is worth noting that the responses of the group delay to human blood temperature and plasma concentration all conformed to cubic functions. In this way, tunable slow light was achieved in this waveguide by varying the temperature of human blood and the concentration of blood plasma.

It should be noted that the largest group delay obtained in this paper was only about 0.1 ps. One reason is that the size of the waveguide was small. The other is that the structural parameters of the waveguide were optimized in order to obtain multiple Fano resonances with high transmittance, failing to take into account the group delay. One can obtain a large group delay by increasing the contrast of Fano resonance and reducing the bandwidth of Fano resonance. Here, only a potential solution for adjusting slow light in the MIM waveguide was proposed, combined with biological parameters.

From current research results, it can be seen that the coupling efficiency between a MIM waveguide and external optical devices is relatively low, due to the small width of the waveguide. If the coupling efficiency can be improved, the application of the MIM waveguide may be greatly promoted.

4. Discussion

Triple Fano resonances were achieved in the MIM waveguide made up of a SS-RCSC, CRCSC, and bus waveguide. The refractive index may be changed to tune the Fano resonances individually. The maximum sensitivity of refractive index sensing was 2453.33 nm/RIU. The cavities of this waveguide can be filled with biological solutions as a biosensor, which make it possible to measure several biological parameters at once. Human blood temperature and plasma concentration sensing had maximal sensitivities of 0.25 nm/°C and 0.2 nm·L/g, respectively. Meanwhile, tunable slow light can be realized using this waveguide, and the group delay responses to human blood temperature and plasma concentration all conformed to cubic functions. The significant advantage of this waveguide is that two separate resonant cavities allow for the simultaneous measurement of multiple biological parameters and slow light tuning of multiple wavelengths. In conclusion, the waveguide proposed in this paper can play a role in biosensing and optical information processing in nanoscale applications.

Author Contributions: H.T.: conceptualization, resources, supervision, funding acquisition, and writing—review and editing. R.Z.: conceptualization, methodology, software, investigation, data curation, and writing—original draft. Y.L. and S.C.: visualization. All authors have read and agreed to the published version of the manuscript.

Funding: This research was funded by Innovation and Entrepreneurship Training Program for Undergraduates of Northeast Forestry University (DC-2023177).

Institutional Review Board Statement: Not applicable.

Informed Consent Statement: Not applicable.

Data Availability Statement: Not applicable.

Conflicts of Interest: The authors declare no conflict of interest.

References

1. Fano, U. Effects of Configuration Interaction on Intensities and Phase Shifts. *Phys. Rev.* **1961**, *124*, 1866–1878. [CrossRef]
2. Dai, X.; Ruan, B.; Xiang, Y. Self-Referenced Refractive Index Biosensing with Graphene Fano Resonance Modes. *Biosensors* **2021**, *11*, 400. [CrossRef]
3. Limonov, M.F.; Rybin, M.V.; Poddubny, A.N.; Kivshar, Y.S. Fano resonances in photonics. *Nat. Photonics* **2017**, *11*, 543–554. [CrossRef]
4. Luk'yanchuk, B.; Zheludev, N.I.; Maier, S.A.; Halas, N.J.; Nordlander, P.; Giessen, H.; Chong, C.T. The Fano resonance in plasmonic nanostructures and metamaterials. *Nat. Mater.* **2010**, *9*, 707–715. [CrossRef]
5. Wan, F.; Qian, G.; Li, R.; Tang, J.; Zhang, T. High sensitivity optical waveguide accelerometer based on Fano resonance. *Appl. Opt.* **2016**, *55*, 6644–6648. [CrossRef]

6. Li, J.; Chen, J.; Liu, X.; Tian, H.; Wang, J.; Cui, J.; Rohimah, S. Optical sensing based on multimode Fano resonances in metal-insulator-metal waveguide systems with X-shaped resonant cavities. *Appl. Opt.* **2021**, *60*, 5312–5319. [CrossRef] [PubMed]
7. Ziauddin, J.R.; Chaung, Y.-L.; Rahmatullah. Tunable Fano resonances via optomechanical effect and gain -loss ratio in coupled microresonators. *Laser Phys.* **2018**, *28*, 116003. [CrossRef]
8. Li, J.; Yu, R.; Wu, Y. Actively tunable double-Fano and Ramsey-Fano resonances in photonic molecules and improved sensing performance. *Phys. Rev. A* **2016**, *94*, 063822. [CrossRef]
9. Deng, Y.; Cao, G.; Yang, H. Tunable Fano resonance and high-sensitivity sensor with high figure of merit in plasmonic coupled cavities. *Photonics Nanostruct. Fundam. Appl.* **2018**, *28*, 45–51. [CrossRef]
10. Liu, G.; Cheng, D.; Zhang, B.; Shu, G.; Wang, J. A microwave biosensor based on spoof surface plasmon polaritons for in vivo measurement of the water content of human skin tissues. *J. Phys. D Appl. Phys.* **2019**, *52*, 205401. [CrossRef]
11. Chen, Z.-Q.; Yin, Z.-X.; Xia, G.-Q.; Hong, L.-L.; Hu, Y.-L.; Liu, M.-H.; Hu, X.-W.; Kudryavtsev, A.A. Pulsed microwave-driven argon plasma jet with distinctive plume patterns resonantly excited by surface plasmon polaritons. *Chin. Phys. B* **2015**, *24*, 025203. [CrossRef]
12. Balevicius, Z. Photonic Sensors in Chemical and Biological Applications. *Biosensors* **2022**, *12*, 1021. [CrossRef] [PubMed]
13. Tong, L.; Wei, H.; Zhang, S.; Xu, H. Recent Advances in Plasmonic Sensors. *Sensors* **2014**, *14*, 7959–7973. [CrossRef]
14. Xia, S.; Zhai, X.; Wang, L.; Wen, S. Plasmonically induced transparency in double-layered graphene nanoribbons. *Photonics Res.* **2018**, *6*, 692–702. [CrossRef]
15. Ruffato, G.; Pasqualotto, E.; Sonato, A.; Zacco, G.; Silvestri, D.; Morpurgo, M.; De Toni, A.; Romanato, F. Implementation and testing of a compact and high-resolution sensing device based on grating-coupled surface plasmon resonance with polarization modulation. *Sens. Actuators B Chem.* **2013**, *185*, 179–187. [CrossRef]
16. Wang, Y.; Jia, S.; Qin, J. Tunable Fano Resonance and Enhanced Sensing in Terahertz Metamaterial. *Front. Phys.* **2021**, *8*, 605125. [CrossRef]
17. Choi, H.; Pile, D.F.P.; Nam, S.; Bartal, G.; Zhang, X. Compressing surface plasmons for nano-scale optical focusing. *Opt. Express* **2009**, *17*, 7519–7524. [CrossRef]
18. Rohimah, S.; Tian, H.; Wang, J.; Chen, J.; Li, J.; Liu, X.; Cui, J.; Hao, Y. Tunable multiple Fano resonances based on a plasmonic metal-insulator-metal structure for nano-sensing and plasma blood sensing applications. *Appl. Opt.* **2022**, *61*, 1275–1283. [CrossRef]
19. Kong, Y.; Cao, J.; Qian, W.; Liu, C.; Wang, S. Multiple Fano Resonance Based Optical Refractive Index Sensor Composed Of Micro-Cavity and Micro-Structure. *IEEE Photonics J.* **2018**, *10*, 6804410. [CrossRef]
20. Zhang, D.; Cheng, L.; Shen, Z. Formation Laws of Direction of Fano Line-Shape in a Ring MIM Plasmonic Waveguide Side-Coupled with a Rectangular Resonator and Nano-Sensing Analysis of Multiple Fano Resonances. *Crystals* **2021**, *11*, 819. [CrossRef]
21. Chen, J.; Li, J.; Liu, X.; Rohimah, S.; Tian, H.; Qi, D. Fano resonance in a MIM waveguide with double symmetric rectangular stubs and its sensing characteristics. *Opt. Commun.* **2021**, *482*, 126563. [CrossRef]
22. Xia, S.; Zhai, X.; Wang, L.; Xiang, Y.; Wen, S. Plasmonically induced transparency in phase-coupled graphene nanoribbons. *Phys. Rev. B* **2022**, *106*, 075401. [CrossRef]
23. Wang, T.-B.; Wen, X.-W.; Yin, C.-P.; Wang, H.-Z. The transmission characteristics of surface plasmon polaritons in ring resonator. *Opt. Express* **2009**, *17*, 24096–24101. [CrossRef]
24. Park, J.; Kim, H.; Lee, B. High order plasmonic Bragg reflection in the metal-insulator-metal waveguide Bragg grating. *Opt. Express* **2008**, *16*, 413–425. [CrossRef]
25. Rohimah, S.; Tian, H.; Wang, J.; Chen, J.; Li, J.; Liu, X.; Cui, J.; Xu, Q.; Hao, Y. Fano Resonance in the Plasmonic Structure of MIM Waveguide with r-Shaped Resonator for Refractive Index Sensor. *Plasmonics* **2022**, *17*, 1681–1689. [CrossRef]
26. Wu, T.; Liu, Y.; Yu, Z.; Peng, Y.; Shu, C.; Ye, H. The sensing characteristics of plasmonic waveguide with a ring resonator. *Opt. Express* **2014**, *22*, 7669–7677. [CrossRef]
27. Zhang, Q.; Huang, X.-G.; Lin, X.-S.; Tao, J.; Jin, X.-P. A subwavelength coupler-type MIM optical filter. *Opt. Express* **2009**, *17*, 7549–7555. [CrossRef]
28. Lin, X.-S.; Huang, X.-G. Tooth-shaped plasmonic waveguide filters with nanometeric sizes. *Opt. Lett.* **2008**, *33*, 2874–2876. [CrossRef]
29. Yun, B.; Hu, G.; Cui, Y. Theoretical analysis of a nanoscale plasmonic filter based on a rectangular metal-insulator-metal waveguide. *J. Phys. D Appl. Phys.* **2010**, *43*, 385102. [CrossRef]
30. Feng, C.; Ying, Z.; Zhao, Z.; Gu, J.; Pan, D.Z.; Chen, R.T. Wavelength-division-multiplexing (WDM)-based integrated electronic-photonic switching network (EPSN) for high-speed data processing and transportation. *Nanophotonics* **2020**, *9*, 4579–4588. [CrossRef]
31. Liu, X.; Li, J.; Chen, J.; Rohimah, S.; Tian, H.; Wang, J. Fano resonance based on D-shaped waveguide structure and its application for human hemoglobin detection. *Appl. Opt.* **2020**, *59*, 6424–6430. [CrossRef]
32. Petrovszki, D.; Krekic, S.; Valkai, S.; Heiner, Z.; Der, A. All-Optical Switching Demonstrated with Photoactive Yellow Protein Films. *Biosensors* **2021**, *11*, 432. [CrossRef]
33. Jiang, C.; Jiang, L.; Yu, H.; Cui, Y.; Li, X.; Chen, G. Fano resonance and slow light in hybrid optomechanics mediated by a two-level system. *Phys. Rev. A* **2017**, *96*, 053821. [CrossRef]

34. Yi, X.; Tian, J.; Yang, R. Tunable Fano resonance in MDM plasmonic waveguide with a T-shaped resonator coupled to ring resonator. *Mater. Res. Express* **2019**, *6*, 035021. [CrossRef]
35. Liu, Y.; Tian, H.; Zhang, X.; Wang, M.; Hao, Y. Quadruple Fano resonances in MIM waveguide structure with ring cavities for multisolution concentration sensing. *Appl. Opt.* **2022**, *61*, 10548–10555. [CrossRef]
36. Liang, Z.; Wen, Y.; Zhang, Z.; Liang, Z.; Xu, Z.; Lin, Y.-S. Plasmonic metamaterial using metal-insulator-metal nanogratings for high-sensitive refraction index sensor. *Results Phys.* **2019**, *15*, 102602. [CrossRef]
37. Rakhshani, M.R.; Mansouri-Birjandi, M.A. High-Sensitivity Plasmonic Sensor Based on Metal-Insulator-Metal Waveguide and Hexagonal-Ring Cavity. *IEEE Sens. J.* **2016**, *16*, 3041–3046. [CrossRef]
38. Chen, J.; Lian, X.; Zhao, M.; Xie, C. Multimode Fano Resonances Sensing Based on a Non-Through MIM Waveguide with a Square Split-Ring Resonance Cavity. *Biosensors* **2022**, *12*, 306. [CrossRef]
39. Qi, J.; Chen, Z.; Chen, J.; Li, Y.; Qiang, W.; Xu, J.; Sun, Q. Independently tunable double Fano resonances in asymmetric MIM waveguide structure. *Opt. Express* **2014**, *22*, 14688–14695. [CrossRef] [PubMed]
40. Fang, Y.; Wen, K.; Qin, Y.; Li, Z.; Wu, B. Multiple fano resonances in an end-coupled MIM waveguide system. *Opt. Commun.* **2019**, *452*, 12–17. [CrossRef]
41. Wang, M.; Tian, H.; Liu, X.; Li, J.; Liu, Y. Multiparameter Sensing Based on Tunable Fano Resonances in MIM Waveguide Structure with Square-Ring and Triangular Cavities. *Photonics* **2022**, *9*, 291. [CrossRef]
42. Duc Duy, N.; Zhao, S. A new high order dispersive FDTD method for Drude material with complex interfaces. *J. Comput. Appl. Math.* **2015**, *285*, 1–14. [CrossRef]
43. Gugger, H.; Jurich, M.; Swalen, J.D.; Sievers, A.J. Reply to "Comment on 'Observation of an index-of-refraction-induced change in the Drude parameters of Ag films'". *Phys. Rev. B Condens. Matter* **1986**, *34*, 1322–1324. [CrossRef] [PubMed]
44. Wen, K.; Chen, L.; Zhou, J.; Lei, L.; Fang, Y. A Plasmonic Chip-Scale Refractive Index Sensor Design Based on Multiple Fano Resonances. *Sensors* **2018**, *18*, 3181. [CrossRef]
45. Dong, S.; Liu, H.; Zheng, Y.; Zhang, J.; Xia, S.; Dong, C.; Shen, K.; Deng, C.; Luo, W.; Su, M.; et al. Numerical study on the biosensing in mid-infrared based on multiple Fano-resonance plasmonic waveguide. *Optik* **2022**, *270*, 170042. [CrossRef]
46. Wang, Q.; Ouyang, Z.; Lin, M.; Liu, Q. Independently Tunable Fano Resonances Based on the Coupled Hetero-Cavities in a Plasmonic MIM System. *Materials* **2018**, *11*, 1675. [CrossRef]
47. Liu, X.; Li, J.; Chen, J.; Rohimah, S.; Tian, H.; Wang, J. Independently tunable triple Fano resonances based on MIM waveguide structure with a semi-ring cavity and its sensing characteristics. *Opt. Express* **2021**, *29*, 20829–20838. [CrossRef]
48. Al-Naib, I. Terahertz Asymmetric S-Shaped Complementary Metasurface Biosensor for Glucose Concentration. *Biosensors* **2022**, *12*, 609. [CrossRef]
49. Zhang, Y.; Kuang, Y.; Zhang, Z.; Tang, Y.; Han, J.; Wang, R.; Cui, J.; Hou, Y.; Liu, W. High-sensitivity refractive index sensors based on Fano resonance in the plasmonic system of splitting ring cavity-coupled MIM waveguide with tooth cavity. *Appl. Phys. A* **2019**, *125*, 13. [CrossRef]
50. Chen, Z.; Cao, X.; Song, X.; Wang, L.; Yu, L. Side-Coupled Cavity-Induced Fano Resonance and Its Application in Nanosensor. *Plasmonics* **2016**, *11*, 307–313. [CrossRef]
51. Xiao, G.; Xu, Y.; Yang, H.; Ou, Z.; Chen, J.; Li, H.; Liu, X.; Zeng, L.; Li, J. High Sensitivity Plasmonic Sensor Based on Fano Resonance with Inverted U-Shaped Resonator. *Sensors* **2021**, *21*, 1164. [CrossRef]
52. Wen, K.; Hu, Y.; Chen, L.; Zhou, J.; He, M.; Lei, L.; Meng, Z.; Wu, Y.; Li, J. Fano Resonance Based on End-Coupled Cascaded-Ring MIM Waveguides Structure. *Plasmonics* **2017**, *12*, 1875–1880. [CrossRef]
53. Rakhshani, M.R. Refractive index sensor based on concentric triple racetrack resonators side-coupled to metal-insulator-metal waveguide for glucose sensing. *J. Opt. Soc. Am. B Opt. Phys.* **2019**, *36*, 2834–2842. [CrossRef]
54. Yahya, M.; Saghir, M.Z. Empirical modelling to predict the refractive index of human blood. *Phys. Med. Biol.* **2016**, *61*, 1405–1415. [CrossRef] [PubMed]
55. El-Khozondar, H.J.; Mahalakshmi, P.; El-Khozondar, R.J.; Ramanujam, N.R.; Amiri, I.S.; Yupapin, P. Design of one dimensional refractive index sensor using ternary photonic crystal waveguide for plasma blood samples applications. *Phys. E Low Dimens. Syst. Nanostruct.* **2019**, *111*, 29–36. [CrossRef]

photonics

MDPI

Article

Range-Gated LIDAR Utilizing a LiNbO3 (LN) Crystal as an Optical Switch

Chenglong Luan [1], Yingchun Li [2,*], Huichao Guo [2] and Houpeng Sun [1]

[1] Graduate School, Space Engineering University, Beijing 101416, China; a1054909975@163.com (C.L.); sunhoupeng202210@163.com (H.S.)

[2] Department of Electronic and Optical Engineering, Space Engineering University, Beijing 101416, China; guohuichao@163.com

* Correspondence: fengfeng521@stu.xjtu.edu.cn; Tel.: +86-176-0100-7310

Abstract: In this paper, a range-gated LIDAR system utilizing an LN crystal as the electro-optical switch and a SCMOS (scientific complementary metal oxide semiconductor) imaging device is designed. To achieve range-gated operations, we utilize two polarizers and an LN (LiNbO3) crystal to form an electro-optical switch. The optical switch is realized by applying a pulse voltage at both ends of the crystal due to the crystal's conoscopic interference effect and electro-optical effect. The advantage of this system is that low-bandwidth detectors, such as a CMOS and a CCD (charge-coupled device), can be used to replace conventional high-bandwidth detectors, such as an ICCD (intensified charge-coupled device), and it displays better imaging performance under specific conditions at the same time. However, after using an electro-optical crystal as an optical switch, a new inhomogeneity error will be introduced due to the conoscopic interference effect of the electro-optical crystal, resulting in a range error for the LIDAR system. To reduce the influence of inhomogeneity error on the system, this paper analyzes the sources of inhomogeneity error caused by the electro-optical crystal and calculates the crystal's inhomogeneity mathematical expression. A compensation method is proposed based on the above inhomogeneity mathematical expression. An experimental LIDAR system is constructed in this paper to verify the validity of the compensation method. The experimental results of the range-gated LIDAR system show that in a specific field of view (2.6 mrad), the LIDAR system has good imaging performance; its ranging standard deviation is 3.86 cm and further decreases to 2.86 cm after compensation, which verifies the accuracy of the compensation method.

Keywords: range-gated; LIDAR; conoscopic interference; electro-optic crystal

1. Introduction

The three-dimensional LIDAR system can be used to obtain the range and three-dimensional image of the target, so it is widely used in space target detection, landscape mapping, underwater target detection, unmanned car driving, and other fields [1]. LIDARs that use focal plane arrays and array detectors can achieve excellent 3D imaging performance, but using these advanced detectors is too costly [2]. In order to maintain some key performance of LIDAR to a certain extent while minimizing cost, many researchers have proposed many different LIDAR schemes to replace the original high-cost detectors [3–14]. Among many 3D imaging LIDAR systems, the use of high-speed optical switches and low-bandwidth, low-cost CCD and CMOS detectors is a common solution to achieve high resolution and high distance accuracy. In 2016, Sungeun Jo et al. in South Korea proposed a high-precision 3D imaging LIDAR system consisting of a polarization-modulating Pockels cell (PMPC) and an MCCD, which can achieve a ranging accuracy of 5.2 mm at 16 m [15]. In 2017, Chen Zhen et al. of the Chinese Academy of Sciences proposed a flash LIDAR system based on polarization modulation using two EMCCD cameras and two PMPCs, which can achieve a ranging accuracy of 0.26 m in a field of view of 0.92 mrad at 200 m [16]. In 2019, Wang Shengjie et al. of the Chinese Academy of Sciences proposed a high-precision correction algorithm for a large field of view based on the LIDAR system proposed by Chen Zhen et al., which reduced the distance error of the system to less than 0.1 m within a 0.92 mrad field of view [17]. In 2020, Song Yishuo et al. from the Space Engineering University proposed a LIDAR system using two KTN crystals and two CCD cameras, which is characterized by a large imaging field of view and a range accuracy of 4.4 cm for targets 15 m away under a field of view of 0.35 rad [18].

In order to meet the needs of 3D imaging and attitude measurement of unmanned aerial vehicles and other targets in various complex environments, ensure the spatial resolution, operating distance, and distance accuracy of the experimental system, and reduce costs, we decided to construct a range-gated active 3D imaging LIDAR experimental system. For commonly used range-gated active 3D imaging LIDARs that use area array imaging devices, ICCD cameras with high-speed shutters are generally used as imaging devices. However, the imaging process of ICCD cameras has several drawbacks, such as multiple electro-optical conversions, low quantum efficiency, low lateral resolution caused by pixel coupling, and a high cost [19,20]. An EMCCD and SCMOS are common alternatives to ensure effective range and spatial resolution. Compared to EMCCD cameras, SCMOS cameras have a higher signal-to-noise ratio, a faster readout speed, and a higher real-time imaging frame rate at the same exposure time, except for inferior imaging quality in extremely weak light. Furthermore, among these three types of cameras, SCMOS cameras have the lowest price. However, SCMOS cameras do not have high-speed shutters and therefore do not have time resolution capabilities. Using SCMOS cameras alone cannot achieve high-precision range-gated imaging. In order to enable the system to achieve high-precision, range-gated imaging, it is necessary to add high-speed optical switches to the system.

In recent years, there has been significant progress in the research on optical switches. In 2012, Long Chen and Young-kai Chen from the USA demonstrated a compact, low-loss, and low-power broadband 8 × 8 optical silicon switch. The optical switching device they demonstrated had a footprint of 8 × 8 mm, and its port-to-port isolation is above 30 dB over the whole 80-nm-wide spectral range and above 45 dB near 30 nm [21]. In 2015, Rafael C. Figueiredo et al. from Brazil presented an ultrafast electro-optical amplified switch based on a chip-on-carrier semiconductor optical amplifier with a high optical contrast of up to 33 dB at a bandwidth of 1550 nm. Switching times up to 115 ps with a small overshoot were achieved by using the multi-impulse step injected current technique [22]. In 2019, Yijian Huang et al. from China reported a liquid-crystal-filled photonic crystal fiber for electro-optical modulation. The device they designed exhibits response and recovery times of approximately 47 ms and 24 ms, respectively. Additionally, the device's operation wavelength can be tuned linearly across a broadband from 1414 nm to more than 1700 nm [23]. In the same year, Kaixuan Chen et al. from China reported a broadband optical switch based on a densely packed silicon waveguide array. The optical switching functionality was realized experimentally with excess losses less than 1.3 dB and crosstalks less than 15 dB over a 60 nm bandwidth for the two spatial modes. An arbitrary splitting ratio at two output ports for the two modes simultaneously was also demonstrated when applying different currents to the heater [24]. In 2021, Giuseppe Brunetti et al. from Italy designed a large-bandwidth 2 × 2 interferometric switching cell. The photonic switch they designed showed a worst-case extinction ratio of approximately 13 dB, insertion loss of less than 2 dB, and crosstalk of 12 dB over a broad bandwidth of 150 nm [25]. In 2023, Jie Tang et al. from China designed a LNOI-based high-speed electro-optical switch. The switch time was less than 13.4 ns, and the extinction ratio was approximately 31.8 dB at a bandwidth of 1550 nm [26].

To meet our needs for a large aperture, high extinction ratio, high switching speed, and relatively low cost, we used the LiNbO3 (LN) crystal and two polarizers placed vertically before and after the crystal as an electro-optical switch. The standard deviation of the range calculated using the original slice image method is 3.86 cm. In comparison, the standard deviation calculated using the slice images compensated by the compensation method is 2.86 cm at a field of view of 2.6 mrad, which verified the validity of the compensation method.

2. Distance-Gating LIDAR Utilizing an Electro-Optical Crystal as the Optical Switch

The schematic diagram of the structure of the range-gated LIDAR system using an electro-optical crystal as an optical switch is shown in Figure 1. The system consists of a SCMOS camera, two polarizers, an electro-optical crystal, a narrow-band filter, a collimator, a receiving optical system, a pulsed laser, a microlens array, a beam expander, a delay signal generator, and a computer. Among them, the polarizer (P1 in Figure 1), the analyzer (P2 in Figure 1), whose polarization direction is perpendicular to the polarizer, and the electro-optical crystal constitute a crystal optical switch. The optical path can be switched on and off by changing the applied voltage of the crystal. The laser is a low-repetition-rate, high-pulse-energy, and narrow-pulse-width pulsed solid-state laser with a wavelength of 532 nm.

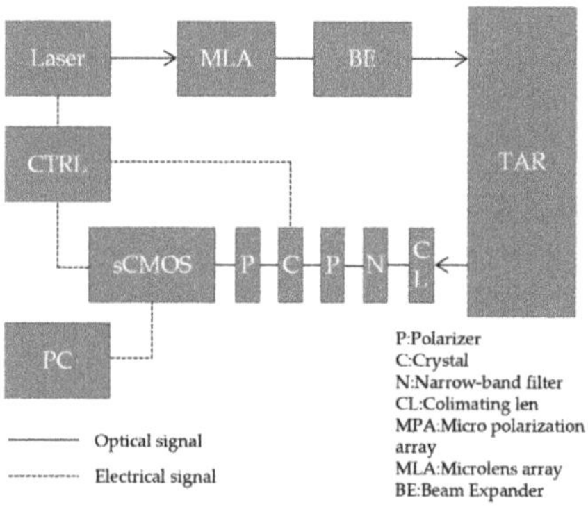

Figure 1. Schematic diagram of the LIDAR system using an electro-optical crystal as an optical switch.

When the LIDAR system is working, a homogeneous and expanded pulse laser beam, whose pulse width is τ_p, is emitted at the target. When the pulse laser travels between the target and the receiving system, the applied voltage of the LN crystal is zero. Currently, the optical switch is in a closed state to prevent atmospheric backscattering and non-target scattering light. After the delay time τ_D, when the reflected light of the target returns to the receiving system, the stray light of other bands is filtered out first by the NBF (narrow-band filter), and then, under the control of the delay signal generator, a pulse voltage with a pulse width of τ_g and a pulse peak of the half-wave voltage of the LN crystal is added at both ends of the crystal.

At this time, the crystal electro-optical switch is on, and the SCMOS camera receives the reflected light of the target. After this process, the system completes the imaging of the target. The corresponding slice depth of field is:

$$\Delta z = \frac{c \cdot (\tau_p + \tau_g)}{2} \tag{1}$$

where c is the speed of light, and we can see from the formula that the crystal optical switch is only turned on for a short time so that the picture obtained by the SCMOS camera corresponds to a certain distance information. A series of distance-slice images can be obtained by changing the delay time τ_D. The distance information of each image pixel can be calculated through the temporal and energy relationships between different slices. Then a three-dimensional image of the target can be obtained [3].

The mainstream ranging algorithm is the weighted average method [10]. This method is to multiply each gray value of the slice image sequence of the same pixel by a delay time serial number weight and divide it by the sum of all serial numbers to obtain an average image serial number. In order to calculate the distance, the average serial number is multiplied by the delay step and the delay time corresponding to the first image; this value is used as the flight time that corresponds to the target distance. This method has the advantages of high precision and fast operational speed.

According to the ranging principle above, the distance corresponding to the ith image in the slice sequence image is:

$$r = \frac{c}{2}\tau_i = \frac{c}{2}(\tau_D + i\Delta\tau) \quad (2)$$

where τ_i represents the delay time corresponding to the ith image, τ_D represents the initial delay time, and $\Delta\tau$ is the delay step. The average image serial numbers can be obtained as follows:

$$< i >= \frac{\sum\limits_i i \times L_i}{\sum\limits_i L_i} \quad (3)$$

where L_i represents the gray value of each pixel in the ith image. The range corresponding to each pixel can be obtained through the following formula:

$$< r >= \frac{c}{2n}(t_0 + < i > \Delta t) \quad (4)$$

When the range of the target point is determined according to the geometric relationship between the target image, the imaging optical system, and the target, the spatial position of the target point can be obtained, and the coordinates of each target point obtained can be converted into three-dimensional coordinates, and then the three-dimensional image of the target can be obtained by display processing.

3. Analysis of Electro-Optic Crystal Switch Inhomogeneity

We used a SCMOS camera to acquire the target image in our LIDAR system. As mentioned above, the SCMOS camera itself does not have a high-speed switch and thus can only obtain grayscale information. This means that it is necessary to use an optical switch that is composed of an electro-optical crystal and two polarizers to achieve slice imaging so that the gray value of the image can correspond to information at a certain distance.

The electro-optic crystal can be used as an optical switch mainly due to its conoscopic interference effect. The normalized light intensity of the central conoscopic interference pattern generated by the light passing through the optical switch is theoretically 0 when the voltage applied to the electro-optic crystal is 0. The electro-optic crystal optical switch can be regarded as being in an "off" state. Theoretically, the normalized light intensity of the central conoscopic interference pattern generated by the light passing through the optical switch is 1 when the voltage applied to the electro-optic crystal is half-wave. The electro-optic crystal optical switch can be regarded as being in an "on" state at this time. Compared with the switch of the ICCD cameras, the distribution of the conoscopic interference ring generated by the electro-optic crystal is not uniform, resulting in a deviation between the gray value of the image generated by the SCMOS cameras with electro-optic crystal switches and the SCMOS cameras with ideal switches. As can be seen from Equation (7) above, an electro-optic crystal switch will cause errors in the image sequence's calculated mean, affecting the LIDAR system's performance. An appropriate field of view needs to be selected to reduce this effect.

We chose an LN crystal as the electro-optic switch, and the inhomogeneity of the LN crystal was analyzed below. Combined with the light tracing analysis of the crystal with reference [27], we can conclude that when the light travels along the crystal optical axis, the crystal phase difference delay caused by the electro-optic effect is:

$$\Gamma = \frac{2\pi}{\lambda} n_o^3 \gamma_{22} \frac{V}{d} L \tag{5}$$

The crystal conoscopic interference light intensity expression is:

$$I(\alpha, \beta, V) = \sin^2[2\phi(\alpha, \beta)] \sin^2[\frac{\theta(\alpha, \beta, V)}{2}] \tag{6}$$

where α is the zenith angle of the light incident to the electro-optical crystal, β is the azimuth angle of the incident light, V is the applied voltage of the crystal, n_o is the refractive index of ordinary light in the crystal, γ_{22} is the electro-optical coefficient of the crystal, and L and d are the length and thickness in the direction of the applied pulse voltage, respectively. It can be seen that the ordinary light refractive index n_o, electro-optical coefficient γ_{22}, length L, and thickness d of the crystal are fixed values for a certain crystal, so the light intensity of the crystal is a function of α, β, and V as variables. The influencing factors of the light intensity inhomogeneity of crystal switches are also mainly derived from α, β, and V.

The influence of light intensity inhomogeneity is divided according to the abovementioned variables. The total inhomogeneity coefficient μ is defined here, which represents the difference between the actual normalized light intensity received by a pixel on the detector and the normalized light intensity of the corresponding pixel in the center of the field of view on the detector, that is, the difference between the transmittance of the crystal optical switch of the pixel in the center of the field of view and the transmittance of the crystal optical switch corresponding to any other pixel. The inhomogeneity coefficient caused by the difference in incidence angles α and β (shown in the Figure 2) is defined as κ, representing the difference in transmittance of the crystal optical switch due to the influence of the conoscopic light interference effect. The inhomogeneity coefficient caused by the unideal pulse voltage V is defined as ρ, representing the difference between the transmittance of the crystal under the actual non-ideal pulse voltage and the ideal pulse voltage. The total inhomogeneity coefficient is the sum of the two factors:

$$\mu = \kappa + \rho \tag{7}$$

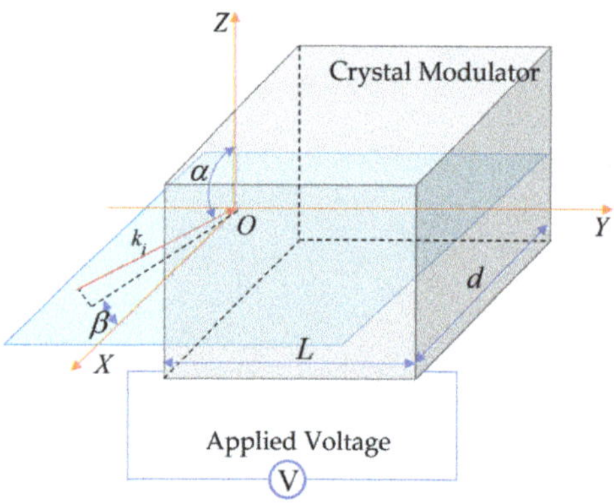

Figure 2. Schematic diagram of the angle of light incident on the crystal surface.

3.1. Analysis of Inhomogeneity Caused by the Angle of Incident Light of the Crystal

Firstly, the inhomogeneity caused by the different angles κ of the crystal's incident light is analyzed. According to Equation (6), the interference pattern of the crystal under ideal conditions is calculated and simulated using MATLAB, and the simulation result is shown in Figure 3.

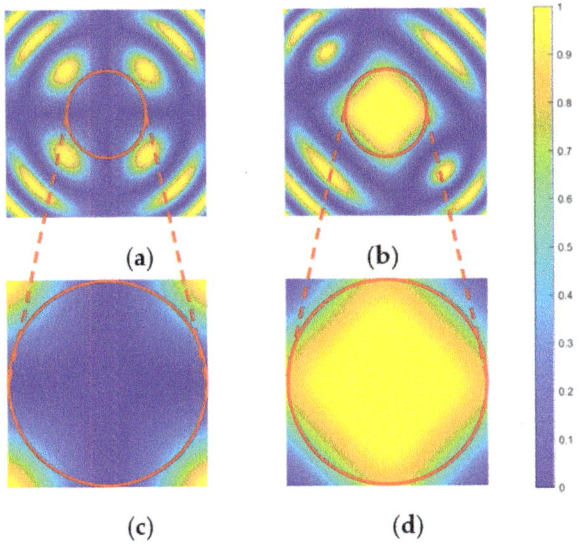

Figure 3. Normalized light intensity patterns received by the detector after the light passes through the electro-optical switch under ideal conditions. (**a**) and (**c**) are the patterns at a 50 mrad field of view angle and a 20 mrad field of view angle when the voltage applied to the crystal is 0, respectively; (**b**) and (**d**) are the patterns of the crystal at a 50 mrad field of view angle and a 20 mrad field of view angle when a half-wave voltage is applied to the crystal, respectively.

As shown in Figure 3, Figure 3a shows the normalized intensity distribution of light passing through the crystal when the voltage V applied across the crystal is 0 and the field of view angle is 50 mrad. Figure 3c is an enlarged view of the field angle range of 20 mrad in the center of Figure 3a. From Figure 3a,c, it can be seen that the uniformity of the image center is good, but the image uniformity is poor in the field of view edge areas at azimuth angles of 45°, 135°, 225°, and 315°. According to the MATLAB simulation results, at the center of the red circle, i.e., the central field of view, the normalized light intensity is 0. At the edge of the red box, which is the 20 mrad field of view angle, the normalized light intensity is 0.0168.

Figure 3b shows the normalized intensity distribution of light passing through the crystal when the voltage at both ends of the crystal is half-wave voltage V_{π} and the field of view is 50 mrad. Figure 3d is an enlarged view of the field angle range of 20 mrad in the center of Figure 3b. According to the simulation results, the normalized light intensity at the center of the field of view is 1, while at the edge of the 20 mrad field of view, the normalized light intensity is only 0.782.

From the simulation results, it can be seen that when LN crystals are used as optical switches, as the field of view angle changes, the normalized intensity of the transmitted light will change, thereby affecting the uniformity of the light intensity received by the detector.

According to the analysis above, based on the normalized light intensity value at the center of the field of view, when the applied voltage is a specific voltage V, the difference between the normalized light intensity received by a pixel of the detector and the normalized light intensity received at the center of the field of view is:

$$\kappa = \sin^2[2\phi(\alpha,\beta)]\left\{1-\sin^2[\frac{\theta(\alpha,\beta,V_\pi)}{2}]\right\} \tag{8}$$

3.2. Analysis of Inhomogeneity Caused by Unideal Pulse Voltage Applied to the Crystal

The inhomogeneity ρ caused by the pulse voltage V applied to the crystal is analyzed below.

In order to calculate the inhomogeneity caused by pulse voltage, the pulse voltage applied to the electro-optical crystal should be measured with an oscilloscope. It can be seen from the measurement results below that the pulse width of the applied voltage source used in this research group is 100 ns.

Combined with Equation (6), the conoscopic interference pattern produced by the crystal is a function of voltage, and the corresponding conoscopic interference pattern is different for different voltages. The light intensity distribution detected on the detector during the process of applying a half-wave pulse voltage to the electro-optical crystal without considering the laser pulse waveform is:

$$I(\alpha,\beta,V_{mean}) = \frac{1}{n}[\sum_{V_i}^{n} I(\alpha,\beta,V)\cdot\eta(i)] \tag{9}$$

Among them, $I(\alpha,\beta,V_{mean})$ represents the light intensity of the image detected by the detector after one exposure,V_i represents the voltage value of the ith sampling point on the oscilloscope, n is the total number of the laser pulse voltage sampling points on the oscilloscope, $I(\alpha,\beta,V_i)$ represents the light intensity corresponding to the crystal conoscopic interference pattern when the voltage applied to the crystal is V_i, and $\eta(i)$ represents the normalized light intensity of the laser pulse received by the crystal when the voltage is V_i. When the ideal electro-optic crystal switch is open, the voltage applied at both ends should be the half-wave voltage of the crystal. As shown by the waveform of the crystal external driving power supply displayed on the oscilloscope in Figure 4, there is a significant error between the crystal external pulse voltage and the ideal pulse half-wave voltage. The non-uniformity coefficient corresponding to the light intensity error caused by the imperfect pulse voltage is calculated with Equation (10).

$$\begin{aligned}\rho &= I(\alpha,\beta,V_\pi) - I(\alpha,\beta,V_{mean})\\ &= \sin^2[2\phi(\alpha,\beta)]\left\{\sin^2[\tfrac{\theta(\alpha,\beta,V_\pi)}{2}]-\sin^2[\tfrac{\theta(\alpha,\beta,V_{mean})}{2}]\right\}\end{aligned} \tag{10}$$

In this formula, $I(\alpha,\beta,V_\pi)$ is the normalized conoscopic interference light intensity corresponding to the half-wave voltage.

Combining the influence of the conoscopic interference effect of the LN crystal and the non-ideal pulse voltage, the total inhomogeneity coefficient generated by the electro-optic crystal optical switch can be obtained by bringing Equations (8) and (10) into Equation (7):

$$\mu = \sin^2[2\phi(\alpha,\beta)]\left\{\sin^2[\frac{\theta(0,0,V)}{2}]+\sin^2[\frac{\theta(\alpha,\beta,V_\pi)}{2}]-\sin^2[\frac{\theta(\alpha,\beta,V)}{2}]-\sin^2[\frac{\theta(\alpha,\beta,V_{mean})}{2}]\right\} \tag{11}$$

When the voltage at both ends of the crystal is half-wave voltage V_π, the formula can be simplified as:

$$\mu = \sin^2[2\phi(\alpha,\beta)]\left\{1-\sin^2[\frac{\theta(\alpha,\beta,V_{mean})}{2}]\right\} \tag{12}$$

Based on the non-uniformity analysis discussed above, we used MATLAB R2021a software to simulate the imaging results of the system.

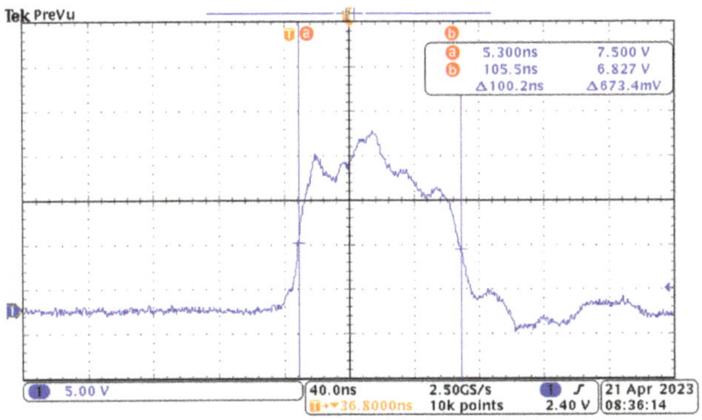

Figure 4. Image of the applied voltage of the crystal detected by the oscilloscope. The a and b mark the vertical cursors of the oscilloscope which measure the time and amplitude of the signal. The measurement results are marked in the first two lines of the legend in the upper right corner while the data in the third line represents the difference between the measurement results of two cursors. The yellow T represents the triggering time of the signal.

Figure 5 shows the temporal relationship between the laser pulse and the voltage applied to the crystal. Figure 6 shows the results of gated imaging of planar targets by a range-gated LIDAR system using an LN crystal as a switch under ideal conditions. The relationship between the pulse voltage applied to the crystal and the laser pulse corresponding to the simulated imaging results in Figure 6a is shown in Figure 5a. Additionally, Figure 6b–i shows the simulation images corresponding to the increase in the gating delay by 4 ns, respectively. The relationship between the pulse voltage applied to the crystal and the laser pulse corresponding to the simulated imaging results in Figure 6i is shown in Figure 5b. Combined with Figure 5, when the laser pulse entirely coincides with the pulse voltage applied to the crystal in a time sequence within a 2.6 mrad field of view, the maximum normalized light intensity in the simulation image Figure 6a is 1, and the minimum normalized light intensity is 0.96, which can be regarded as the case where the plane target is completely gated. With the increase in delay time, the coincidence region of the laser pulse and the pulse voltage applied to the crystal gradually decrease, and only part of the laser pulse is gated. The normalized light intensity of the image decreases from both ends to the center, and the normalized light intensity of the image's upper-left and lower-right ends is significantly higher than that of the lower-left and upper-right ends. The simulation results reflect the apparent inhomogeneity of the system slice image.

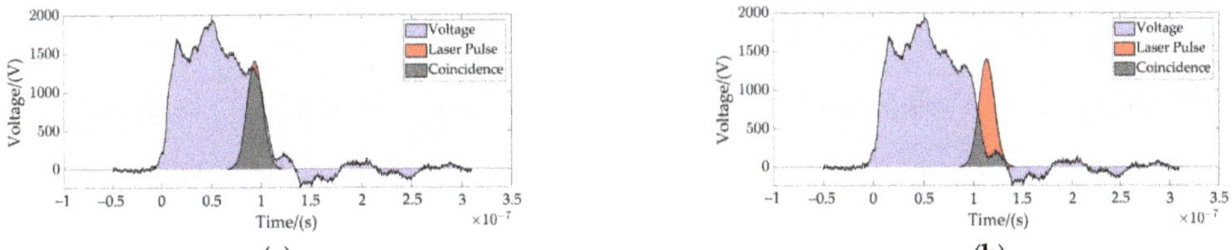

Figure 5. Temporal relationship between the laser pulse and the voltage applied to the crystal. The purple region represents the laser pulse voltage, the orange region represents the laser pulse, and the gray region represents the coincidence region of the laser pulse and voltage applied to the crystal. (**a**) indicates that the laser pulse and the applied pulse voltage are completely coincident. (**b**) indicates that the laser pulse partially leaves the applied pulse voltage of the crystal.

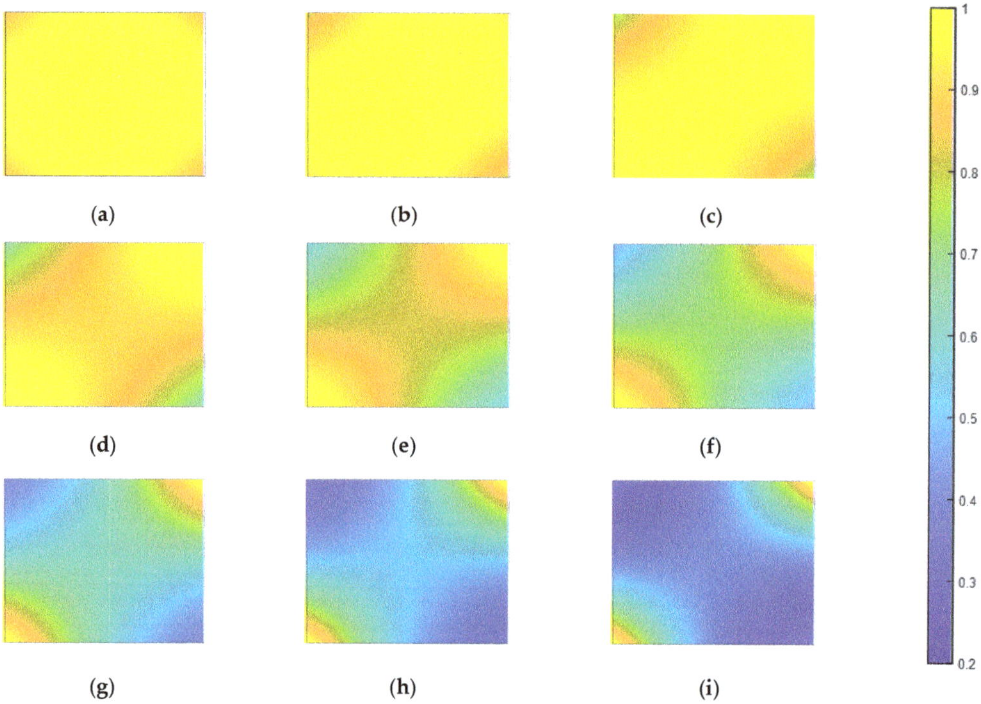

Figure 6. MATLAB simulation results of the plane target imaging within a 2.6 mrad field of view. Among them, the sequential relationship between the applied pulse voltage and laser pulse corresponding to the simulated imaging results in (**a**) is shown in Figure 5a, and the sequential relationship between the crystal applied pulse voltage and the laser pulse corresponding to the simulated imaging result in (**i**) is shown in Figure 5b. (**b–i**) is the simulation image corresponding to the increase in the gating delay by 4 ns, respectively.

3.3. A Compensation Method for Crystal Inhomogeneity

According to the above analysis of the source of electro-optical crystal switch inhomogeneity, the compensation method for crystal inhomogeneity is studied. In the actual compensation process, not only the relationship between the crystal ideal matrix, the inhomogeneity matrix, and the compensation matrix should be considered, but also a certain extinction ratio ε of the switch should be taken into consideration as the actual switch is not ideal.

We define the transmittance matrix represented by $\mathbf{Z}$ as the full α matrix, and α represents the transmittance of the ideal switch under a certain voltage applied to the crystal. The actual electro-optic crystal switch transmittance matrix is defined as $\mathbf{Z}'$, and the transmittance of the center field of view in $\mathbf{Z}'$ should also be α. The inhomogeneity matrix calculated by Formula (12) is defined as μ, the actual inhomogeneity matrix considering the extinction ratio is defined as μ', the compensation matrix is $\mathbf{C}$, and the extinction ratio matrix is defined as ε. Then μ' can be calculated using the following equation:

$$\mu' = \left(\mu + \frac{\mathbf{A}}{\varepsilon\text{-}\mathbf{A}}\right) / \mathrm{M}\left(\mu + \frac{\mathbf{A}}{\varepsilon\text{-}\mathbf{A}}\right) \quad (13)$$

In the formula above, **A** represents a matrix with all values of 1, and M is the maximum value function that is used to normalize the matrix. After obtaining the actual inhomogeneity matrix, the compensation matrix **C** can be calculated using the following formula:

$$\mathbf{C}=\frac{\mathbf{Z}}{\mathbf{Z}'}=\frac{\mathbf{Z}}{\mathbf{Z}-\mu} \tag{14}$$

For the slice image captured by the experimental LIDAR system, the image compensating the inhomogeneity of the electro-optic crystal switch can be obtained by multiplying pixel by pixel with the corresponding compensation matrix. It should be noted here that according to Equation (1), the slice image obtained has a certain depth of field, meaning that a slice image may contain laser pulses reflected by targets at different distances. The time of flight of the laser pulse reflected from different distances is different, and the applied pulse voltage of the crystal corresponding to the laser pulse received by the system at different times is also different, resulting in different transmittances of the electro-optic crystal switch. When the pulse width of the pulse voltage applied to the crystal and the laser pulse width of the crystal are large, the interference state of the electro-optic crystal switch corresponding to the target at different distances on a single slice image is quite different. At this time, when the compensation matrix corresponding to a specific time is applied to the slice image, a more significant error might be introduced. The compensation matrix can only be used when the laser pulse width and the pulse width of the voltage applied to the crystal are small to a certain extent.

4. Experiment and Analysis

Table 1 shows the main parameters of the experimental LIDAR system. The experimental system of the proposed range-gated experimental LIDAR using an LN crystal as an optical switch is shown in Figure 7. Figure 8a is the optical layout of the experimental system. As shown in Figures 7 and 8a, we use a Tektronix's AFG 31,000 SERIES arbitrary signal generator (ASG) and a Fastlaser.Tech's TDG-VII timing delay generator (TDG) to realize the synchronous control of the crystal voltage driver, SCMOS camera exposure control, laser pulse xenon lamp signal, and Q-switched signal. Since the ASG we use produces up to two signals, the TDG produces a delay of up to 1 ms, which is less than the exposure delay of the PCO Edge.55 SCMOS camera, and the TDG's jitter time in the external trigger mode is much smaller than the corresponding jitter time when using the internal clock, we chose to combine the signal generator and the signal delayer to achieve synchronization control of the entire experimental system.

Table 1. Main parameters of the range-gated experimental LIDAR system using an LN crystal as an optical switch.

Parameter	Value
Wavelength	532 nm
Pulse Energy	200 mJ
Pulse Width	7 ns
Pulse Repetition Frequency	5 Hz
Aperture	400 mm
Detector	2160 × 2560 pixels

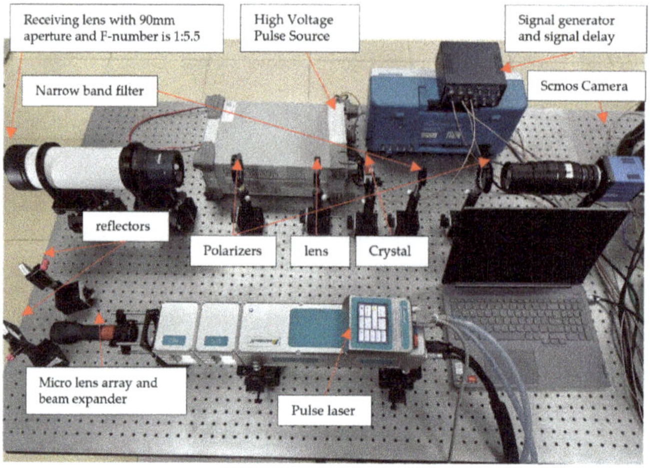

Figure 7. Range-gated experimental LIDAR system using an LN crystal as an optical switch.

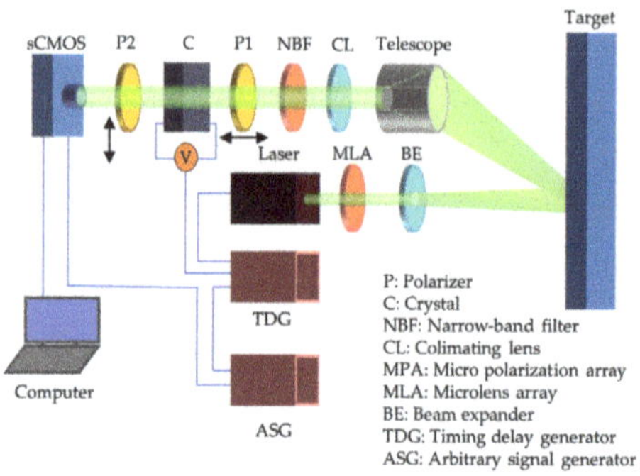

(**a**)

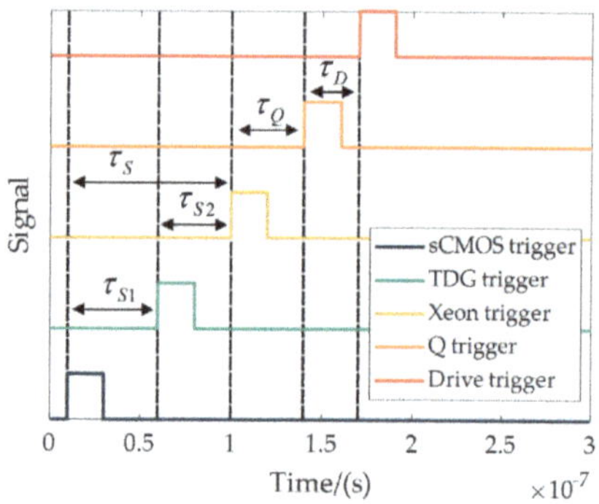

(**b**)

Figure 8. (**a**) Optical layout of the experimental system; (**b**) Sequential chart of the experimental system.

As shown in Figure 8a, when the system is working, the ASG generates a signal without delay and a signal with a delay of τ_{S1} to trigger the SCMOS camera and the TDG, respectively. After the TDG is triggered, the trigger signal generated by the TDG triggers the laser xenon lamp, the Q-switched switch, and the driver of the LN crystal. The delay τ_Q between the xenon lamp trigger signal and the Q-switched trigger signal is the Q-switched delay of the laser, and the delay τ_D between the Q-switched trigger signal and the driver of the LN crystal trigger signal is the range-gated delay. As shown in Figure 8b, when the laser is triggered, a pulsed xenon lamp-pumped Nd:YAG Q-switched laser emits a pulsed laser with a pulse width of 7 ns and a spot distribution similar to a Gaussian distribution at the target. The pulsed laser first reaches the microlens array and is homogenized, then reaches the beam expander and is expanded by it. After that, the pulsed laser propagates to the target and is reflected by the target. The reflected laser pulse first reaches the telescope with an aperture of 90 mm and an F number of 1:5.5, and after being received by the telescope, it reaches the collimator CL and is collimated. The collimated laser pulse then reaches the NBF and filters out the stray light of other wavelengths, thereby improving the signal-to-noise ratio of the system. After passing through the NBF, the laser pulse reaches an optical switch composed of polarizer P1 with a horizontal polarization direction, an LN crystal C with a length and width of 9 mm and a thickness of 18 mm, and polarizer P2 with a vertical polarization direction. If the crystal optical switch is closed under the action of the driving voltage at this time, the pulsed laser is blocked by the optical switch, and the SCMOS cannot receive the laser pulse. If the crystal optical switch is turned on under the action of the pulse voltage at this time, the laser pulse is received by the SCMOS camera, thus enabling a range-gated image of the target.

In order to verify the accuracy of the compensation method, a pure white plate is placed 22.04 m away from the receiving system, and a gated imaging experiment on the white plate is conducted. The effect of imaging compensation is shown in Figure 9. From Figure 9a, we can see that because the physical principle of the crystal as an optical switch is the conoscopic interference effect of the crystal, only the area of about 301 × 301 pixels in the center of the field of view can achieve uniform imaging. Figure 9c,d are the contrast images of the imaging effect of the 601 × 601 pixel range area of the crystal center before and after compensation. From the image effect before and after compensation, it can be seen that in the pixel area taken, the gray image uniformity of the target imaging is significantly improved, proving the algorithm's accuracy.

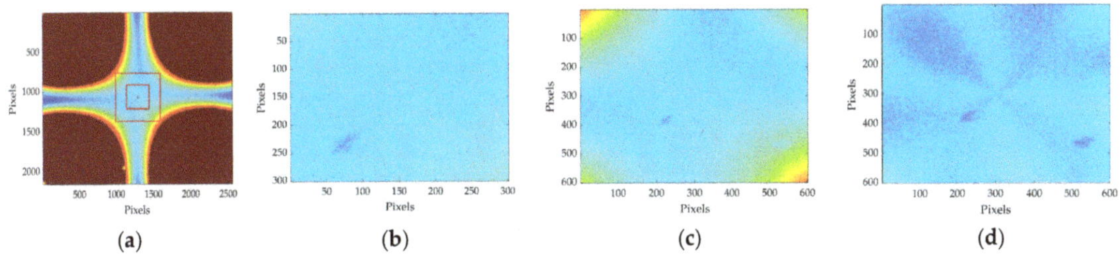

Figure 9. (**a**) The gray image corresponding to the whiteboard's ungated imaging. (**b**) The 301 × 301 pixel enlarged image before compensation. (**c**) The 601 × 601 pixel enlarged image in the central field of view. (**d**) The 601 × 601 pixel enlarged image in the central field of view after compensation.

Using the slice image before compensation and the slice image after compensation for three-dimensional reconstruction, the obtained point cloud images areshown in Figure 10.

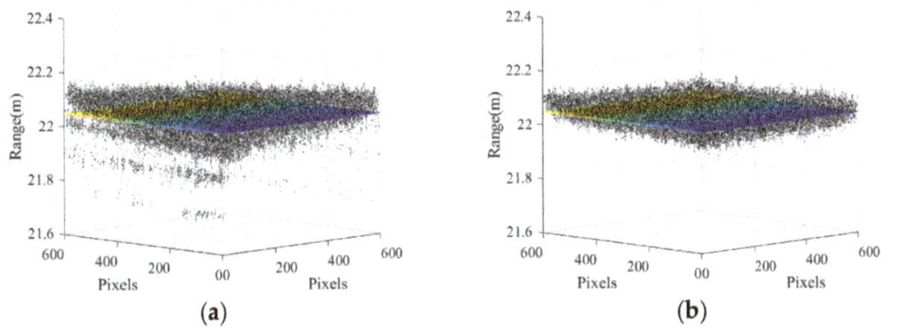

Figure 10. Three-dimensional reconstruction point cloud image before and after compensation; the colorful plane is the plane of the actual plate. (**a**) Point cloud recovered from the slice image obtained by using the system to directly gate the white plate; (**b**) Point cloud recovered from the compensated slice image.

Figure 11 are the distance depth maps of the flat target obtained by the system before and after compensation. Figure 12 shows the comparison of the distance depth with the horizontal row of pixels, where the orange curve represents the distance depth curve in Figure 11b, the green curve represents the distance depth curve in Figure 11a, and the blue line represents the true distance of the target. Figure 12a–c represent the average distance depth of the first to fifth rows of pixels in Figure 11, the average distance depth of the 299th to 303rd row of pixels, and the average distance depth of the 597th to 601st rows of pixels, and these pixels are marked with solid black translucent lines, respectively, in Figure 12a,b. From Figures 10–12, we can conclude that due to the influence of the inhomogeneity of the electro-optical crystal switch, the calculated value of the distance depth is generally smaller than the true value in the upper-left and lower-right parts of the distance depth image calculated by the original slice image, while the calculated value of the distance depth is generally larger than the true value in the lower-left and upper-right parts of the image. Additionally, there are obvious isolated points at the edge of the point cloud image. The range accuracy of the depth image restored by the compensated slice image is significantly improved in the corner part; the tilt phenomenon is obviously suppressed; the isolated points of the image in the edge part are significantly reduced; and the image dispersion in the center part is reduced. This result verifies the accuracy of the compensation method.

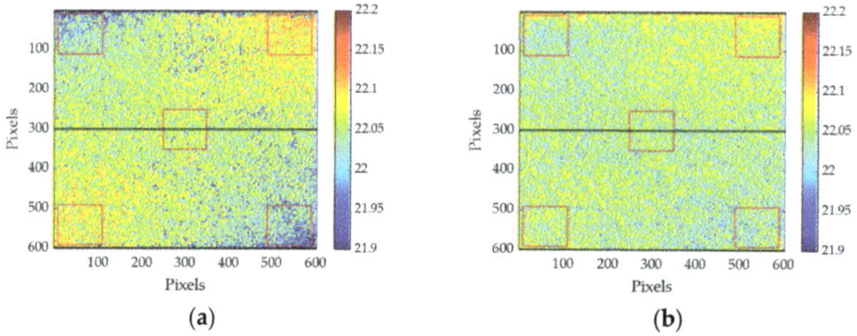

Figure 11. Distance depth map of the flat target obtained by the system before and after compensation. (**a**) Depth map obtained by using the slice image produced after using the system to directly gate the white plate; (**b**) Depth map obtained from the compensated slice image.

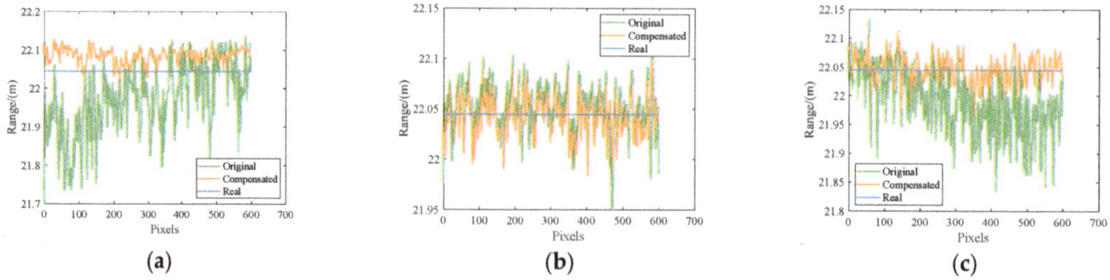

Figure 12. Comparison of horizontal pixel distance depth. (**a–c**) represent the average distance depth of the first to fifth rows of pixels in Figure 11, the average distance depth of the 299th to 303rd row of pixels, and the average distance depth of the 597th to 601st rows of pixels.

The distance accuracy of the image within a 2.6 mrad field of view is calculated as follows:

$$\sigma = \sqrt{\frac{1}{n-1}\sum_{i=1}^{n}(r_i - \bar{r})^2} \quad (15)$$

In the formula, σ represents the standard deviation of distance depth, n represents the total number of pixels, r_i represents the distance corresponding to the ith pixel, and $\bar{r}$ represents the actual measured target distance with a fixed value of 22.04 m. After calculation, the standard deviations of the distance depth corresponding to Figure 11a,b are 3.86 cm and 2.86 cm, respectively. Additionally, the standard deviations of the distance depth of the areas marked with red boxes in the upper-left, upper-right, lower-left, lower-right, and central field of view in Figure 11a are 4.26 cm, 3.18 cm, 2.71 cm, 5.05 cm, and 2.69 cm, respectively, and in Figure 11b, the standard deviations are 2.53 cm, 2.38 cm, 2.56 cm, 2.81 cm, and 2.30 cm, respectively. Among them, the standard deviation of the distance depth of the marked area in the upper-left and lower-right parts decreased significantly, indicating that the tilt phenomenon caused by the crystal switch was significantly suppressed, which verified the accuracy of the compensation method. The leading causes of range error are as follows: First, the pulse voltage applied to the crystal is not an ideal pulse, and its peak waveform stability is poor. In this case, there will be errors when the sequence image weighted average method is used to recover the distance depth. Second, the principle of the crystal as an optical switch is the conoscopic interference effect of the crystal. The conoscopic light interference effect itself has inhomogeneity, and there will be a significant error at the edge of the field of view. Third, the jitter time of the TDG triggering the crystal pulse voltage is 25 ps according to the product manual, and the corresponding ranging error is 0.38 cm. The shot noise of SCMOS will also cause a particular distance error. After compensation, the influence of the first and second reasons is obviously suppressed, and the standard deviation of system distance depth is obviously improved.

5. Conclusions

In this paper, a range-gated LIDAR system using an electro-optic crystal as the optical switch is designed. The slice image of the target is obtained by synchronously controlling the pulse voltage applied to the crystal, laser pulse, and SCMOS exposure, and then the three-dimensional point cloud image and distance depth image of the target are reconstructed using the slice image.

Compared with the traditional range-gated LIDAR system using an ICCD as an imaging device, a high-precision sequential logic circuit and high-bandwidth detector must be used to improve the range accuracy. For the system designed in this paper, the use of the SCMOS detector only needs to detect the intensity of the reflected laser of the target instead of detecting the time of flight (TOF) and does not require a high-precision sequential logic circuit, which greatly reduces the cost of the system.

However, compared with the built-in shutter ICCD, the use of an electro-optic crystal as a light shutter will introduce new inhomogeneity errors, which will affect the ranging accuracy of the system. To reduce the impact of this error, this paper quantitatively analyzes the source of the crystal inhomogeneity error, proposes a compensation method for crystal inhomogeneity, and gives the mathematical expression of the inhomogeneity compensation matrix. The experimental results show that compared with the range depth image restored by the original slice image, the accuracy of the compensated range depth image is significantly improved, and the standard deviation of the range depth is reduced from 3.86 cm to 2.86 cm within a 2.6 mrad field of view, which verifies the accuracy of the algorithm and provides a guarantee for further improving the performance of the system.

Author Contributions: Conceptualization, C.L. and H.G.; methodology, C.L. and Y.L.; software, C.L. and H.S.; validation, H.S.; formal analysis, C.L.; investigation, C.L. and Y.L.; resources, H.G.; data curation, Y.L., writing—original draft preparation, C.L.; writing—review and editing, Y.L. and H.G. All authors have read and agreed to the published version of the manuscript.

Funding: This research received no external funding.

Institutional Review Board Statement: Not applicable.

Informed Consent Statement: Not applicable.

Data Availability Statement: Not applicable.

Conflicts of Interest: The authors declare no conflict of interest.

References

1. Park, J.; Cho, J.; Lee, S.; Bak, S.; Kim, Y. An Automotive LiDAR Performance Test Method in Dynamic Driving Conditions. *Sensors* **2023**, *23*, 3892. [CrossRef] [PubMed]
2. Mcmanamon, P.F.; Banks, P.; Beck, J.; Huntington, A.S.; Watson, E.A. A comparison flash lidar detector options. *Opt. Eng.* **2017**, *56*, 031223. [CrossRef]
3. Monnin, D.; Schneider, A.L.; Christnacher, F.; Lutz, Y. A 3D Outdoor Scene Scanner Based on a Night-Vision Range-Gated Active Imaging System. In Proceedings of the Third International Symposium on 3D Data Processing, Visualization, and Transmission, Chapel Hill, NC, USA, 14–16 June 2006; pp. 938–945.
4. Sun, L.; Sun, D.; Sun, C.; Sun, C.; Sun, H. 3D Range-gated Imaging Method Based on Parabolic Envelope Inversion. *Acta Armamentarii* **2022**, *43*, 1868–1873.
5. Yang, X.; Hao, L.; Wang, Y. Adjustable higher SNR and long-range 3D-imaging cluster lidar based on a coded full-waveform technique. *Appl. Opt.* **2019**, *58*, 4671–4677. [CrossRef] [PubMed]
6. Dong, Y.; Zhu, Z.; Tian, X.; Qiu, L.; Ba, D. Frequency-Modulated Continuous-Wave LIDAR and 3D Imaging by Using Linear Frequency Modulation Based on Injection Locking. *J. Light. Technol.* **2021**, *39*, 2275–2280. [CrossRef]
7. Busck, J.; Heiselberg, H. Gated viewing and high-accuracy three-dimensional laser radar. *Appl. Opt.* **2004**, *43*, 4705–4710. [CrossRef]
8. Zhang, P.; Du, X.; Zhao, J.; Song, Y.; Chen, H. High resolution flash three-dimensional LIDAR systems based on polarization modulation. *Appl. Opt.* **2017**, *56*, 3889–3894. [CrossRef]
9. Yun, J.; Gao, C.; Zhu, S.; Sun, C.; He, H.; Feng, L.; Dong, L.; Niu, L. High-peak-power, single-mode, nanosecond pulsed, all-fiber laser for high resolution 3D imaging LIDAR system. *Chin. Opt. Lett.* **2012**, *10*, 121402.
10. Laurenzis, M.; Christnacher, F.; Monnin, D. Long-range three-dimensional active imaging with superresolution depth mapping. *Opt. Lett.* **2007**, *32*, 3146–3148. [CrossRef]
11. Zhang, X.; Yan, H.; Zhou, Q. Overcoming the shot-noise limitation of three-dimensional active imaging. *Opt. Lett.* **2011**, *36*, 1434–1436. [CrossRef]
12. Redman, B.; Ruff, W.C.; Stann, B.L. Photon Counting, Chirped AM LADAR System and Related Methods. U.S. Patent 7,675,610, 2010.
13. Zhang, Z.; Zhang, J.; Wu, L.; Zhang, Y.; Zhao, Y.; Su, J. Photon-counting chirped amplitude modulation lidar using a smart premixing method. *Opt. Lett.* **2013**, *38*, 4389–4392. [CrossRef] [PubMed]
14. Andersen, J.F.; Busck, J.; Heiselberg, H. Pulsed Raman fiber laser and multispectral imaging in three dimensions. *Appl. Opt.* **2006**, *45*, 6198–6204. [CrossRef] [PubMed]
15. Jo, S.; Hong, J.K.; Bang, H.; Kim, J.W.; Choi, S. High resolution three-dimensional flash LIDAR system using a polarization modulating Pockels cell and a micro-polarizer CCD camera. *Opt. Express* **2016**, *24*, A1580–A1585. [CrossRef]
16. Chen, Z. Research on Three-Dimensional Active Imaging with Polarization-Modulated Method. Ph.D. Thesis, University of Chinese Academy of Sciences, Xi'an, China, May 2017.
17. Wang, S.; Liu, B.; Chen, Z.; Li, H. High-Precision Calibration Algorithm for Large Field-of-View Polarization-Modulated 3D Imaging. *IEEE Photonics Technol. Lett.* **2019**, *31*, 1064–1067. [CrossRef]
18. Song, Y.; Zhao, J.; Wang, B.; Wang, X.; Zhang, J. The potassium tantalate niobate (KTN) crystal-based polarization-modulated 3D ladar with a large field of view. *Opt. Lett.* **2020**, *45*, 5319–5322. [CrossRef] [PubMed]
19. Denvir, D.J.; Conroy, E. Electron-multiplying CCD: The new ICCD. In Proceedings of the International Symposium on Optical Science and Technology, Seattle, WA, USA, 7–11 July 2002.
20. Djazovski, O.; Daigle, O.; Laurin, D.; Bedirian, M.; Ducharme, M.E.; Artigau, É.; Doyon, R. Electron-multiplying CCDs for future space instruments. In Proceedings of the Photonics North, Ottawa, ON, Canada, 3–5 June 2013.
21. Long, C.; Chen, Y.K. Compact, low-loss and low-power 8 × 8 broadband silicon optical switch. *Opt. Express* **2012**, *20*, 18977–18985.
22. Figueiredo, R.C.; Ribeiro, N.S.; Ribeiro, A.M.O.; Gallep, C.M.; Conforti, E. Hundred-Picoseconds Electro-Optical Switching with Semiconductor Optical Amplifiers Using Multi-Impulse Step Injection Current. *J. Light. Technol.* **2015**, *33*, 69–77. [CrossRef]
23. Huang, Y.; Wang, Y.; Zhang, L.; Shao, Y.; Zhang, F.; Liao, C.; Wang, Y. Tunable electro-optical modulator based on a photonic crystal fiber selectively filled with liquid crystal. *J. Light. Technol.* **2019**, *37*, 1903–1908. [CrossRef]
24. Chen, K.; Yan, J.; He, S.; Liu, L. Broadband optical switch for multiple spatial modes based on a silicon densely packed waveguide array. *Opt. Lett.* **2019**, *44*, 907–910. [CrossRef]
25. Brunetti, G.; Marocco, G.; Di Benedetto, A.; Giorgio, A.; Armenise, M.N.; Ciminelli, C. Design of a large bandwidth 2 × 2 interferometric switching cell based on a sub-wavelength grating. *J. Opt.* **2021**, *23*, 085801. [CrossRef]
26. Tang, J.; Wang, Z.; Qian, K.; He, X.; Pan, S.; Wang, C.; Gu, X.; Qian, G.; Kong, Y.; Cheng, T. Design and Fabrication of LNOI Based High-speed Electrooptic Switch. *Optoelectron. Technol.* **2023**, *43*, 7–17.
27. Sun, H.; Li, Y.; Guo, H.; Luan, C.; Zhang, L.; Zheng, H. Electro-optic modulation aberration correction algorithm based on phase difference compensation. *Appl. Opt.* **2022**, *61*, 8982–8987. [CrossRef] [PubMed]

photonics

Communication

A Low-Cost Modulated Laser-Based Imaging System Using Square Ring Laser Illumination for Depressing Underwater Backscatter

Yansheng Hao [1], Yaoyao Yuan [1,*], Hongman Zhang [1], Shao Zhang [2] and Ze Zhang [1,2,3,4]

[1] Qilu Aerospace Information Research Institute, Jinan 250100, China; zhangze@aircas.ac.cn (Z.Z.)
[2] College of Photonics and Optical Engineering, Aerospace Information Technology University, Jinan 250299, China
[3] Aerospace Information Research Institute, Chinese Academy of Sciences, Beijing 100094, China
[4] Key Laboratory of Computational Optical Imaging Technology, Chinese Academy of Sciences, Beijing 100094, China
* Correspondence: yuanyy@aircas.ac.cn

Abstract: Underwater vision data facilitate a variety of underwater operations, including underwater ecosystem monitoring, topographical mapping, mariculture, and marine resource exploration. Conventional laser-based underwater imaging systems with complex system architecture rely on high-cost laser systems with high power, and software-based methods can not enrich the physical information captured by cameras. In this manuscript, a low-cost modulated laser-based imaging system is proposed with a spot in the shape of a square ring to eliminate the overlap between the illumination light path and the imaging path, which could reduce the negative effect of backscatter on the imaging process and enhance imaging quality. The imaging system is able to achieve underwater imaging at long distance (e.g., 10 m) with turbidity in the range of 2.49 to 7.82 NTUs, and the adjustable divergence angle of the laser tubes enables the flexibility of the proposed system to image on the basis of application requirements, such as the overall view or partial detail information of targets. Compared with a conventional underwater imaging camera (DS-2XC6244F, Hikvision, Hangzhou, China), the developed system could provide better imaging performance regarding visual effects and quantitative evaluation (e.g., UCIQUE and IE). Through integration with the CycleGAN-based method, the imaging results can be further improved, with the UCIQUE increased by 0.4. The proposed low-cost imaging system with a compact system structure and low consumption of energy could be equipped with platforms, such as underwater robots and AUVs, to facilitate real-world underwater applications.

Keywords: underwater optical imaging; laser imaging; square ring spot; long distance

1. Introduction

The rapid development of unmanned underwater vehicles (UUVs) increases the activity capacity of human beings to explore the underwater environment, and visual information is essential for performing a number of underwater operations, such as underwater ecosystem monitoring [1,2], topographical mapping [3–5], mariculture [6–8], and marine resource exploration [9,10]. The selective absorption properties of water, such as the great absorption of red light and good maintenance of green and blue light, result in the color distortion of captured underwater images. Moreover, the particles suspended in water reflect the light from the sun and artificial light sources, which can introduce scattering to the image sensor and degrade the captured images. The underwater optical imaging process is described in Figure 1. The camera receives direct components, forward scattering, and backward scattering during the imaging process. The direct component refers to light that travels directly from the light source (e.g., the sun) to the target and then to the camera without scattering, which is the basis for forming the image of the item being taken. Forward scattering means the scattering of light at small angles traveling in the water. Backscattering happens when light is scattered by particles in the water and redirected to the camera, lowering the quality of the captured image; most underwater imaging solutions try to reduce the negative effects of backscattering [11].

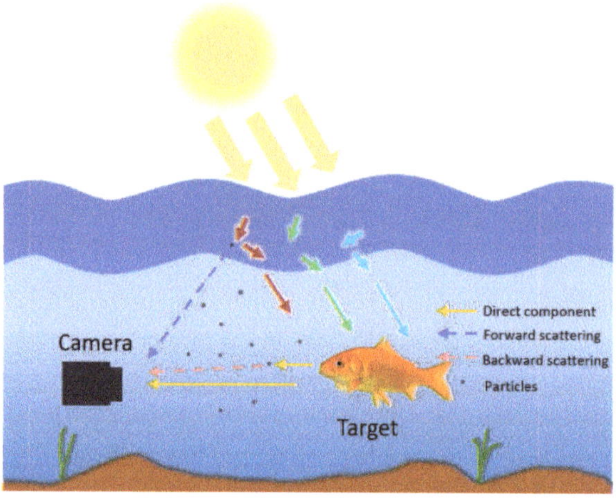

Figure 1. Schematics of the underwater optical imaging process.

Current underwater vision technologies can be grouped into hardware-based methods and software-based methods regarding their principles of operation. Hardware-based methods mainly focus on the improvement of the physical process of imaging to increase the information captured by cameras, while software-based methods rely on the enhancement and restoration of information contained in the images with well-established algorithms. The utilization of lasers in hardware-based imaging systems enhances the ability to obtain high-resolution images of objects and representative laser-enhanced underwater imaging methods, including range-gated imaging, structured light imaging, polarimetric imaging and laser scanning methods, etc. Herein, we mainly discuss the range-gated imaging and laser scanning methods; detailed information about other laser-based methods can be found in other literature sources [12]. The range-gated imaging system employs a synchronized controlled pulsed laser and gate camera with a small gate width to reduce backward scattering for high-resolution imaging. To further remove the backward scattering in gated images, depth-noise maps were calculated with a water attenuation coefficient and a reference image, which were then subtracted with target gate images to obtain a new gate image containing less noise to achieve high range resolution and accuracy 3D imaging [13]. Moreover, it is possible to lessen backscattering by analyzing scene depth with a single gated image, combined with parameters of time delay, laser pulse width, and gate pulse width for dehazing with a 134.78% improvement in PSNR (Peak Signal-to-Noise Ratio) in the underwater environment [14]. Laser scanning methods can capture high-resolution underwater images by reducing the common volume of the light path between the light source and camera to eliminate backscatter. A laser field synchronous scanning system was able to achieve underwater imaging in a range of 15 m and a CSNR (signal-to-noise ratio) improvement of 1.67 times compared with an LED-based imaging system [15]. The performance of the abovementioned range-gated imaging and laser scanning methods is affected by laser parameters, such as laser power and pulse width, which cannot separate the illumination and imaging paths. A range-gated imaging system with high power energy and short exposure time can image long-distance objects at the cost of bulky architecture and high consumption of power. In addition, the field of view (FOV) and imaging resolution of laser scanning methods are restricted by the scanning mirror; moreover, the motion in the imaging process can lead to the twist of high-density point clouds and degrade imaging accuracy.

Software-based methods refer to the methods for improving raw underwater image quality using developed algorithms, which can be classified into imaging enhancement methods, imaging restoration methods, and deep learning-based methods. Imaging enhancement methods process raw underwater images with different functions for advancing the visual effect of the image without considering the imaging physical model; in contrast, imaging restoration methods restore the underwater images by reconstructing the physical imaging formation model in the underwater environment. Typical imaging enhancement methods include Retinex and Fusion-based methods. Retinex aims to restore intrinsic images by eliminating the effects of illumination, which can achieve single underwater image enhancement using multicolor gradient priors of reflectance and illumination to complete specific underwater applications, such as underwater keypoint detection, underwater saliency detection, and underwater depth map estimation [16]. Fusion-based methods are able to reduce noise, expose dark regions, and enhance the contrast of raw images without the assistance of hardware and underwater environment parameters. The improvement of the fusion pipeline, such as reducing the color cast of input images with white balance processing, could enhance the attenuation features and edge information of raw images [17]. Imaging restoration methods try to employ the inverse operation of the underwater imaging formation model to restore a high-quality image with an accurate estimation of model parameters [18]. Deep learning methods, such as CNN and GAN-based methods, use established datasets and networks to learn the information in the images and produce images with good visual results [19,20]. CNN-based methods can learn input image features and provide required outputs, such as transmission maps and image formation model parameters, which can be integrated with the abovementioned enhancement and restoration methods to dehaze the degraded underwater images [21]. GAN-based networks with a generator generate unidentifiable fake images using input images to deceive the discriminator, and a discriminator tries to distinguish the fake images from real images. Due to the lack of datasets with paired images, CycleGAN was proposed to enhance underwater images without the requirement for paired images [22]. However, the performance of these three methods is highly dependent on the input images and accurate estimation of the optical model and parameters, restricting their generalization abilities and the complex network architecture of deep learning-based methods demanding high computability and making it difficult to develop image dehazing methods compatible with different underwater conditions.

In this study, a low-cost and low-power consumption modulated laser-based underwater imaging method was developed with a compact system architecture and long-range and high-quality imaging ability. This system could reduce the backscatter by modulating the laser to match the field of view to improve the SNR (signal-to-noise ratio); moreover, the intensive distribution of laser power in the form of a square ring could expand the imaging range to about 10 m with water turbidity of 7.82 NTUs in starlight illumination conditions. What is more, the performance of this system could be further improved using the CycleGAN-based method, with an obvious improvement in UCIQUE (underwater color image quality evaluation) and IE (information entropy) at different imaging distances. It is convenient to adjust the field of view (FOV) and divergence angle of the modulated laser to image the overall view and partial details of objects in the distance.

2. Methods

2.1. Modulated Laser-Based Imaging System (MLIS)

In order to reduce backscatter, a square-ring modulated laser illumination system was established that matched the rectangular imaging field of view angle through beam modulation. By utilizing the forward scattered light of the laser to illuminate the imaging field of view, the interference of the backward scattered light in the imaging process could be reduced and enhance the imaging quality and the imaging distance in the underwater environment. The light source in the system adopted low-cost semiconductor green laser tubes (wavelength of 532 nm) with special lenses to modulate the spot into a linear laser beam. Compared to a traditional dispersion illumination light with a circular beam, the square ring laser produces a linear beam and the energy is more concentrated, which can effectively reduce the overlap between the illumination light path and the imaging field of view, avoiding interference with backward scattered light on imaging to improve imaging quality, and the dispersion angle of the laser beam can be adjusted to match the field of view to obtain different information regarding the target and surrounding environment for different applications. The camera (JZC-N81820S with a resolution 1920 × 1080, frame rate 25 fps, minimum illumination 0.01 Lux, Xiongmaitech, Hangzhou, China) was placed in the center of the laser illumination source, consisting of a low-light imaging detector and a zoom imaging lens, which can collect target information light under different imaging fields of view. The modulated laser system can be found in Figure 2.

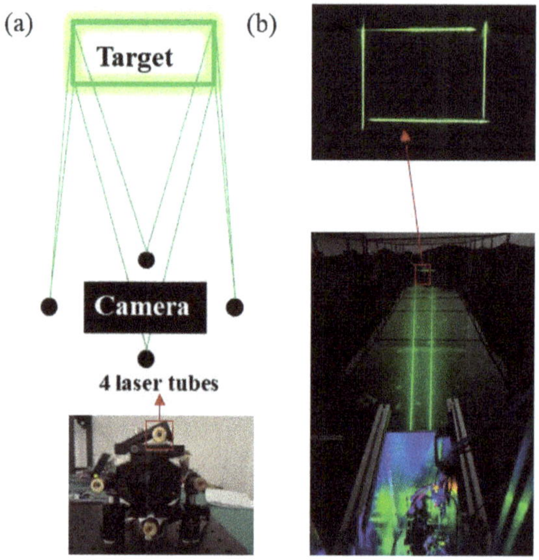

Figure 2. (**a**) Schematics (**top**) and actual figure (**bottom**) of the modulated laser illumination system for underwater imaging. (**b**) Underwater experiment field figure of the modulated laser illumination system for underwater imaging (**bottom**) and square ring laser spot (**top**).

2.2. Electrical Control System

The system mainly consists of a lighting and imaging system and an embedded control system (see Figure 3a). The lighting and imaging system included 2 groups of driver circuits to drive 4 laser tubes with a wavelength of 532 nm (connected in series), 4 shaping lenses, a low illumination imaging module, and an imaging lens. The control system consisted of a microcontroller module, a signal amplification module, a ethernet module, and a power module. The PWM signal was generated and amplified to control the driver circuits for realizing the adjustment of the optical power of the 4 laser tubes. The ethernet module is responsible for signal transmission between the microcontroller and the external device and transmits the data of the imaging module to the external device for further analysis. STM32f103c8t6 with rich peripheral functions such as a timer and USART was used as the embedded controller in this study. With a 32-bit Cortex-M3 core and main frequency up to 72 MHz, this microcontroller realized the balance of high performance and low power consumption, meeting the underwater application requirement of low cost and low power consumption. The diagram and actual figure of the electrical control system based on STM32 can be found in Figure 3b.

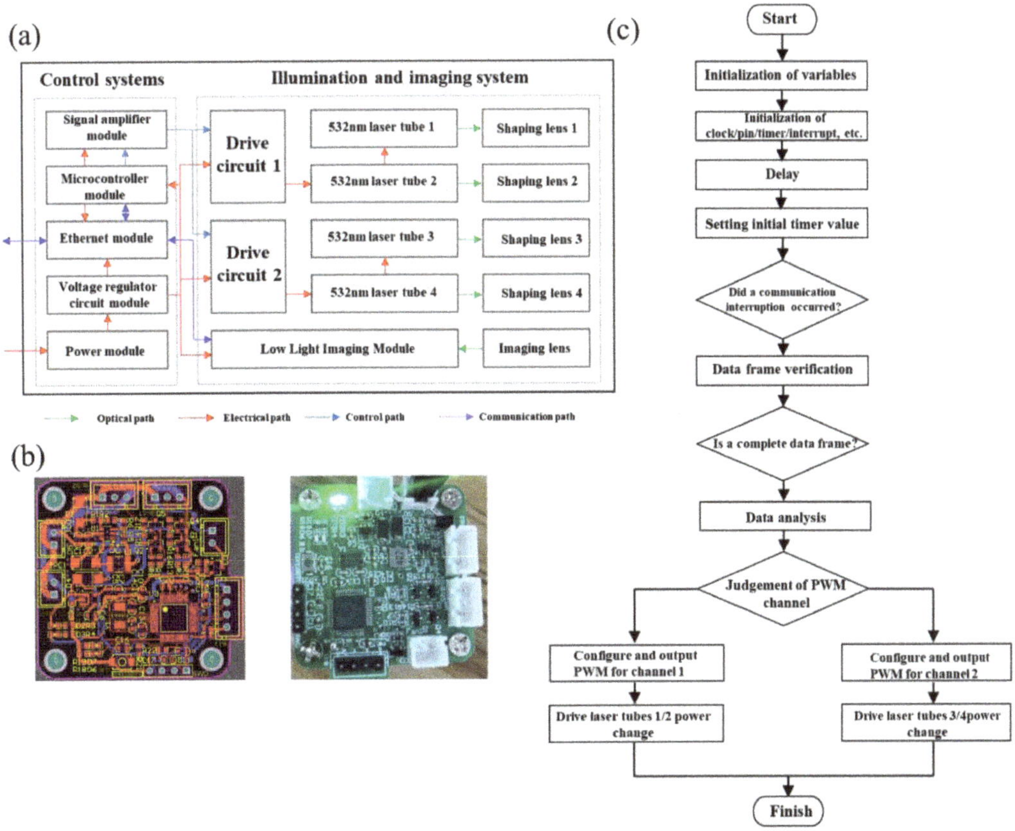

Figure 3. (**a**) The block chain of the optoelectronic system. (**b**) The diagram (**left**) and actual figure (**right**) of the electrical control system based on STM32. (**c**) Flow chart of dedicated firmware.

The embedded programming was designed using Keil MDK5. The PWM was used to modulate the constant-current driver power supply of the laser source, which in turn realized the modulation of the outgoing optical power. PWM regulation was carried out through USART communication, and the baud rate of USART communication was set to 115,200. The controller used a timer to output two PWM signals to control the two driver circuits; each driver circuit controlled two laser tubes. The crossover coefficient of this timer was set to 9 and the duty cycle of the output PWM was changed by adjusting the value of the registers. A detailed flowchart of the firmware can be found in Figure 3c. The overall power consumption of the system was about 9.26 W when the laser illumination was turned on and 3.26 W without illumination; therefore, the overall power consumption was low and suitable for underwater long-time applications. The data stream was mainly generated by the imaging module with a frame rate of 24 fps and an image size of 1920 × 1080 and transmitted to the external device through the ethernet module with an overall data bandwidth of about 142.38 MB/s.

2.3. Imaging Quality Evaluation Metrics

In this study, we UCIQE and IE (information entropy) to quantitatively evaluate the imaging performance of the MLIS. UCIQE is a linear combination of color concentration, saturation, and contrast that is a commonly used metric to carry out quantitative evaluation of underwater images in terms of non-uniform color cast, blurring, and low contrast.

UCIQE is a no-reference (ground truth) metric expressed as Equation (1), where c_1, c_2, and c_3 are weighting factors, σ_c is the standard deviation of chromaticity, con_l is the luminance contrast, and μ_s is the saturation mean.

$$\text{UCIQUE} = c_1\sigma_c + c_2 con_l + c_3\mu_s \quad (1)$$

IE (information entropy) can describe the complexity and richness of detail of a given image, which can be calculated with Equation (2), where L is the number of possible gray levels of the image and $p(i)$ is the probability of the ith gray level appearing in the image, which can be calculated by dividing the total pixel value of the image with the pixel value of the ith gray level.

$$\text{E} = -\sum_{i=0}^{L-1} p(i)\log_2 p(i) \quad (2)$$

3. Experiments

A water pool with dimensions of 10 m × 2 m × 1 m (length × width × depth) was established to simulate the underwater environment. The modulated laser illumination system and camera were packaged in a sealed waterproof compartment and fixed at a depth of 0.5 m under the water surface on one end of the pool. The target was a standard calibration board with a size of 520 mm × 400 mm attached to a movable signpost. In order to remove the effects of sunlight on the imaging results, all of the experiments were performed at night or with the water tank covered with blackout fabric to ensure imaging quality.

The turbidity of the underwater environment could be adjusted in the range of 2.49 to 7.82 NTUs by adding milk to imitate real underwater environment scattering because the milk contained different sizes of spherical particles (e.g., casein molecules of 0.04–0.3 μm, fat globular molecules of 1–20 μm) [23].

Different experiments were carried out to evaluate the performance of the MLIS, including a comparison experiment of modulated laser and diverging laser, the matching effect of FOV and modulated laser divergence angle (MLDA), a comparison experiment of the MLIS with conventional underwater imaging camera (DS-2XC6244F), enhancement with CycleGAN-based method.

4. Results and Discussion

4.1. Comparison of the Modulated Laser and the Diverging Laser

To analyze the effects of the modulated laser on imaging, an underwater experiment was performed to compare the imaging results with the illumination of the modulated laser and diverging laser. The experimental conditions were controlled with water turbidity of 7.82 NTUs and a camera FOV of 15.658° × 8.845°. The choice of laser power should consider the imaging environment (e.g., distance), and low laser energy might not illuminate the target and instead degrade the imaging results. The laser power of 35.8 mW at a distance of 8 m was almost unable to illuminate the target during testing. Therefore, the modulated laser and diverging laser power were 118 mW in the measurement to ensure the illumination of the target.

The results show that modulated laser illumination could provide images with good detail preservation inside the target, uniform image brightness, and clear details of the target and its surrounding background. Under diverging laser illumination, the illumination path overlapped with the imaging path, resulting in a large amount of backscattered light entering the imaging field of view, meaning that the acquired image was mixed with the diverging illumination component in the background with uneven brightness, especially for remote imaging, as can be seen in Figure 4, where the bottom of the target and the surrounding environment were blurred. This finding demonstrates that the modulated laser illumination in the shape of a square ring was suitable for long-range underwater imaging with reasonable adjustment of FOV and modulated laser divergence angle, which will be investigated in the next section. This proposed system only evaluated the imaging distance of 10 m with the water tank length of 10 m; we believe that the imaging distance could be further extended with suitable experimental conditions as the laser at 10 m could still illuminate the target clearly and remain stable, which will be investigated in a future study.

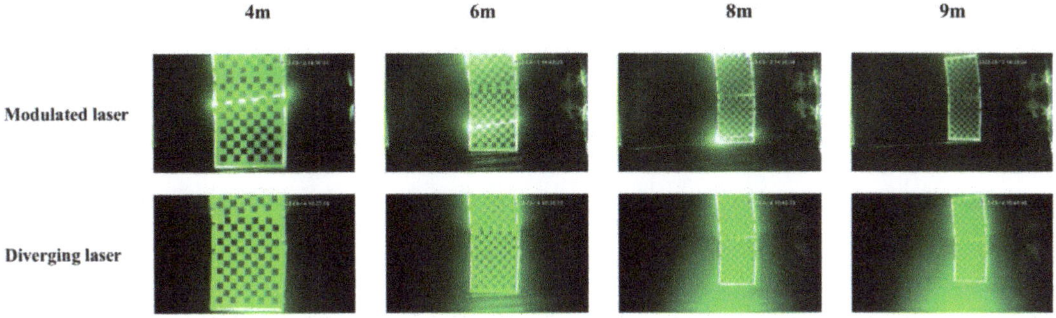

Figure 4. Comparison of original images captured by the camera with the illumination of the modulated laser (**top**) and the diverging laser (**bottom**) at different distances.

4.2. Matching Effect of FOV and Modulated Laser Divergence Angle (MLDA)

This section investigates the matching effect of FOV and MLDA with the FOV fixed as $20.96° \times 11.7°$ and the MLDA adjusted to be smaller than the FOV ($10.01° \times 5.15°$), matched with the FOV ($20.96° \times 11.7°$), and larger than the FOV ($31.28° \times 18.55°$). During the experiments, the water turbidity was 7.82 NTUs, the imaging resolution was 1920×1280, and the power of the modulated laser was 118 mW. If the FOV was smaller than the MLDA, a lot of illuminated information was missed for long-distance imaging, such as imaging with distances of 9 m. When the FOV was larger than the MLDA, the target was imaged with sufficient detail in terms of subjective vision effect but the backscatter of incident light was introduced into the image, resulting in the accumulation of noise in the background. The matched FOV and MLDA preserved the information of the target scene and removed the background noise well (middle row in Figure 5), enabling the performance of the MLIS. What is more, the results implied that it was possible to choose the suitable relationship between the FOV and MLDA based on the application requirements to obtain the required imaging information.

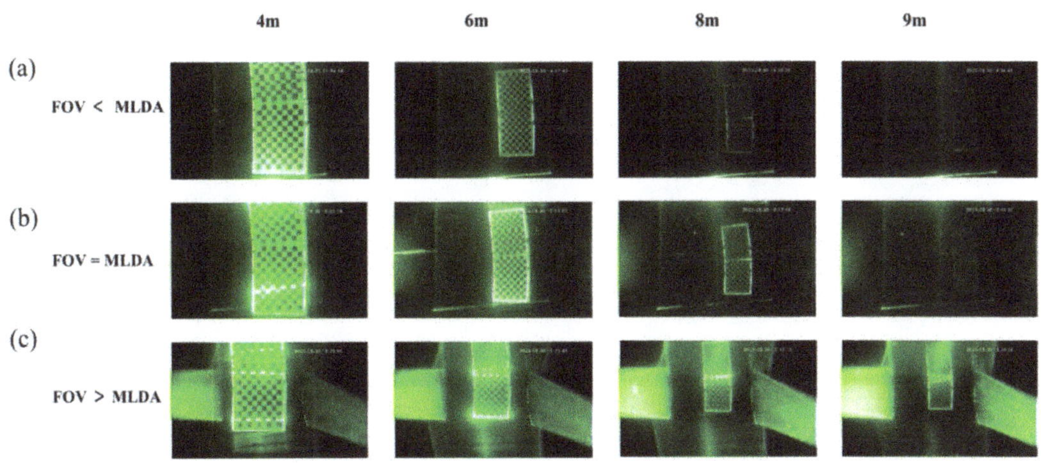

Figure 5. Effects of the relationship between the FOV and MLDA on imaging: (**a**) FOV < MLDA, (**b**) FOV = MLDA, and (**c**) FOV > MLDA.

4.3. Comparison of the MLIS with a Conventional Underwater Imaging Camera (DS-2XC6244F)

The imaging performance of our imaging system and Hikvision product were compared at different imaging distances from 2 m to 10 m under the same imaging conditions (water turbidity of 7.82 NTU, FOV of 8.4° × 4.9°, and resolution of 1920 × 1080). As shown in Figure 6, the conventional underwater imaging camera (DS-2XC6244F, Hikvision, Hangzhou, China) could image the target with the largest distance of 7 m in poor imaging quality and nothing could be identified from the image captured with a distance of 8 m. In comparison, the images captured by the MLIS provided better visual effects for all distances. It is even able to see the edge of the calibration board at the imaging distance of 10 m. To quantitatively evaluate the imaging performance of our imaging system and conventional underwater imaging camera, the IE of images obtained at each distance was calculated and the results can be found in Table 1. It was obvious that each IE value of the images captured by MLIS was higher than that of the conventional underwater imaging camera (DS-2XC6244F). Please note the IE value of the image obtained at 10 m by our system was higher than the IE value at short distance; this might have resulted from the captured background information and would not have affected the evaluation of imaging capacity.

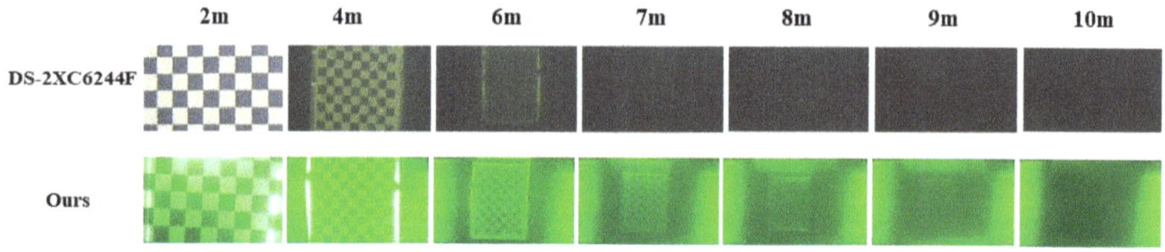

Figure 6. Comparison of original images captured by DS-2XC6244F and the MLIS at different distances.

Table 1. IE of a conventional underwater imaging camera (DS-2XC6244F) and the MLIS.

Methods	2 m	4 m	6 m	7 m	8 m	9 m	10 m
DS-2XC6244F	15.588	13.815	9.41	6.923	6.589	6.861	6.946
MLIS	18.253	15.263	11.287	9.4	8.078	7.67	9.295

4.4. Enhancement with the CycleGAN-Based Method

Through the comparison with the conventional underwater imaging camera (DS-2XC6244F), it was possible to further improve the imaging quality of the MLIS with compatible algorithms. In this study, we chose the CycleGAN-based method to enhance the raw underwater images captured by the MLIS without demanding paired images [24]. The enhanced results in Figure 7 show that the CycleGAN-based method could greatly improve the raw image quality with outstanding visual effects, and the average UCIQUE improved by 0.497. The targets, such as hippocampus japonicas, a starfish, and a crab, could be recognized clearly and the background seaweed could also be displayed vividly, implying the potential of an MLIS integrated with the CycleGAN-based method for a number of underwater applications, such as target recognition and mariculture.

To quantitively assess the enhanced results, the UCIQUE and IE of raw and enhanced images were analyzed, with the results shown in Tables 2 and 3. The evaluation results revealed that the UCIQUE of images captured by the MLIS could be greatly enhanced by the CycleGAN-based method with an average improvement of 0.4. Especially for images captured at long distances, such as 7 m and 10 m, the enhanced UCIQE could improve them by about 0.7 to 0.9. The comparison of IE could give direct results about the complexity and richness of details in the captured images and the enhanced method could greatly improve the IE of images captured using the MLIS. The IE of images captured at distances of 8 m, 9 m, and 10 m could be improved by 4.091, 3.622, and 3.125, respectively. The quantitative comparison results of UCIQUE and IE verified the underwater imaging ability of the proposed system and provided a possible way to improve the performance of MLISs by integrating them with advanced deep learning-based methods. What is more, the obvious improvement of UCIQE and IE for images captured at long distances with the CycleGAN-based method provides the possibility for remote detection with the proposed MLIS for a variety of underwater activities requiring adequate imaging distances for operation, such as underwater topography surveying with a large FOV enabled by an AUV.

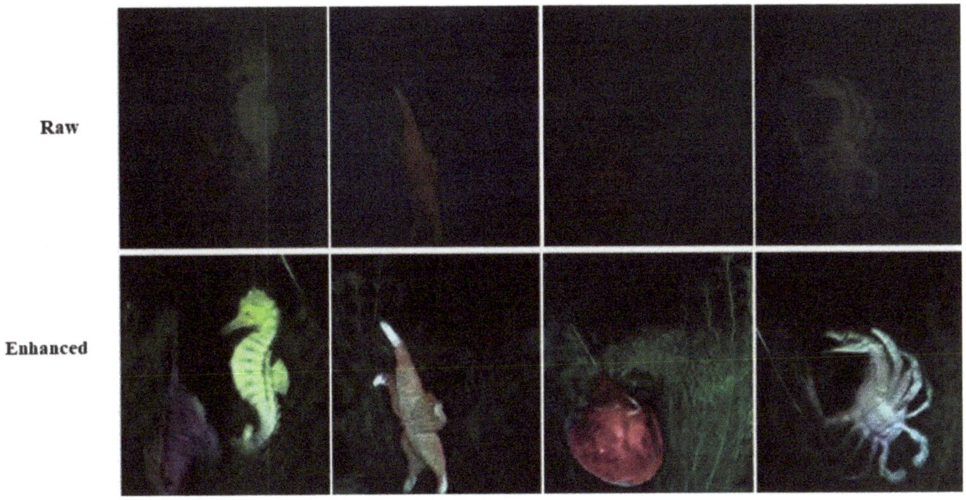

Figure 7. Comparison of images captured by the MLIS and enhanced with the optimized algorithm with the average UCIQUE improved from 0.428 to 0.925.

Table 2. UCIQUE of the MLIS and enhanced MLIS.

Methods	2 m	4 m	6 m	7 m	8 m	9 m	10 m
MLIS	0.408	0.364	0.451	0.477	0.481	0.468	0.538
Enhanced MLIS	0.768	0.637	0.693	1.195	0.895	0.898	1.099

Table 3. IE of the MLIS and enhanced MLIS.

Methods	2 m	4 m	6 m	7 m	8 m	9 m	10 m
MLIS	18.253	15.263	11.287	9.4	8.078	7.67	9.295
Enhanced MLIS	18.192	16.269	13.745	11.414	12.169	11.292	12.42

5. Conclusions

In this manuscript, a modulated laser-based imaging system is proposed with the spot adjusted into a linear laser beam in the shape of a square ring to eliminate the overlap between the illumination light path and the imaging path to reduce the negative effect of backscatter on the imaging process. This imaging system integrated with low-cost laser tubes and cameras provides a method for underwater imaging with long-distance demands (e.g., 10 m) and the adjustable MLDA satisfies the imaging requirement of different scenes for obtaining the overall view or partial details of the target. Compared with a conventional underwater imaging camera (DS-2XC6244F), the developed system can provide better imaging performance regarding visual effects and quantitative evaluation (e.g., UCIQUE and IE); moreover, it is possible to further improve system performance by integrating it with the CycleGAN-based method, with UCIQUE increasing by 0.4. The proposed system could be improved with well-designed imaging enhancement algorithms in the future and we hope that this system could be equipped with a platform, such as an underwater robot and AUV, to facilitate real-world underwater applications.

Author Contributions: Y.H. performed the experiments and wrote the manuscript. Y.Y. proposed the idea and performed the experiments. H.Z. established the optical system and performed the experiments. S.Z. performed the experiments. Z.Z. supervised the manuscript. All authors have read and agreed to the published version of the manuscript.

Funding: This manuscript was funded by the National Natural Science Foundation of China (Grant No. 12304323).

Data Availability Statement: Upon reasonable request, the corresponding author might provide the data supporting the research findings.

Conflicts of Interest: The authors declare no competing financial interests.

References

1. Aguzzi, J.; Thomsen, L.; Flögel, S.; Robinson, N.J.; Picardi, G.; Chatzievangelou, D.; Bahamon, N.; Stefanni, S.; Grinyó, J.; Fanelli, E.; et al. New technologies for monitoring and upscaling marine ecosystem restoration in deep-sea environments. *Engineering* **2024**, *34*, 195–211. [CrossRef]
2. Mariani, P.; Quincoces, I.; Haugholt, K.H.; Chardard, Y.; Visser, A.W.; Yates, C.; Piccinno, G.; Reali, G.; Risholm, P.; Thielemann, J.T. Range-gated imaging system for underwater monitoring in ocean environment. *Sustainability* **2018**, *11*, 162. [CrossRef]
3. Noguchi, Y.; Sekimori, Y.; Maki, T. Guidance method of underwater vehicle for rugged seafloor observation in close proximity. *J. Field Robot.* **2024**, *41*, 314–326. [CrossRef]
4. Leng, Z.; Zhang, J.; Ma, Y.; Zhang, J. Underwater Topography Inversion in Liaodong Shoal Based on GRU Deep Learning Model. *Remote Sens.* **2020**, *12*, 4068. [CrossRef]
5. Mandlburger, G. A review of active and passive optical methods in hydrography. *Int. Hydrogr. Rev.* **2022**, *28*, 8–52. [CrossRef]
6. Darodes de Tailly, J.B.; Keitel, J.; Owen, M.A.; Alcaraz-Calero, J.M.; Alexander, M.E.; Sloman, K.A. Monitoring methods of feeding behaviour to answer key questions in penaeid shrimp feeding. *Rev. Aquac.* **2021**, *13*, 1828–1843. [CrossRef]
7. Wang, Y.; Xiaoning, Y.; An, D.; Wei, Y. Underwater image enhancement and marine snow removal for fishery based on integrated dual-channel neural network. *Comput. Electron. Agric.* **2021**, *186*, 106182. [CrossRef]
8. Liu, X.; Wang, Z.; Yang, X.; Liu, Y.; Liu, B.; Zhang, J.; Gao, K.; Meng, D.; Ding, Y. Mapping China's offshore mariculture based on dense time-series optical and radar data. *Int. J. Digit. Earth* **2022**, *15*, 1326–1349. [CrossRef]
9. Raveendran, S.; Patil, M.D.; Birajdar, G.K. Underwater image enhancement: A comprehensive review, recent trends, challenges and applications. *Artif. Intell. Rev.* **2021**, *54*, 5413–5467. [CrossRef]
10. Chemisky, B.; Menna, F.; Nocerino, E.; Drap, P. Underwater Survey for Oil and Gas Industry: A Review of Close Range Optical Methods. *Remote Sens.* **2021**, *13*, 2789. [CrossRef]
11. Hao, Y.; Yuan, Y.; Zhang, H.; Zhang, Z. Underwater Optical Imaging: Methods, Applications and Perspectives. *Remote Sens.* **2024**, *16*, 3773. [CrossRef]
12. Shen, Y.; Zhao, C.; Liu, Y.; Wang, S.; Huang, F. Underwater optical imaging: Key technologies and applications review. *IEEE Access* **2021**, *9*, 85500–85514. [CrossRef]
13. Wang, M.; Wang, X.; Sun, L.; Yang, Y.; Zhou, Y. Underwater 3D deblurring-gated range-intensity correlation imaging. *Opt. Lett.* **2020**, *45*, 1455–1458. [CrossRef] [PubMed]
14. Wang, M.; Wang, X.; Zhang, Y.; Sun, L.; Lei, P.; Yang, Y.; Chen, J.; He, J.; Zhou, Y. Range-intensity-profile prior dehazing method for underwater range-gated imaging. *Opt. Express* **2021**, *29*, 7630–7640. [CrossRef] [PubMed]
15. Wu, H.; Liu, Z.; Li, C.; Wang, H.; Zhai, Y.; Dong, L. A laser field synchronous scanning imaging system for underwater long-range detection. *Opt. Laser Technol.* **2024**, *175*, 110849. [CrossRef]
16. Zhuang, P.; Li, C.; Wu, J. Bayesian retinex underwater image enhancement. *Eng. Appl. Artif. Intell.* **2021**, *101*, 104171. [CrossRef]
17. Ancuti, C.O.; Ancuti, C.; De Vleeschouwer, C.; Bekaert, P. Color balance and fusion for underwater image enhancement. *IEEE Trans. Image Process.* **2017**, *27*, 379–393. [CrossRef]
18. Zhou, J.; Yang, T.; Chu, W.; Zhang, W.-S. Underwater image restoration via backscatter pixel prior and color compensation. *Eng. Appl. Artif. Intell.* **2022**, *111*, 104785. [CrossRef]
19. Menon, A.; Aarthi, R. A Hybrid Approach for Underwater Image Enhancement using CNN and GAN. *Int. J. Adv. Comput. Sci. Appl.* **2023**, *14*, 742–748. [CrossRef]
20. Cong, R.; Yang, W.; Zhang, W.; Li, C.; Guo, C.L.; Huang, Q.; Kwong, S. PUGAN: Physical Model-Guided Underwater Image Enhancement Using GAN with Dual-Discriminators. *IEEE Trans. Image Process.* **2023**, *32*, 4472–4485. [CrossRef]
21. Zhou, J.; Yang, T.; Zhang, W. Underwater vision enhancement technologies: A comprehensive review, challenges, and recent trends. *Appl. Intell.* **2023**, *53*, 3594–3621. [CrossRef]
22. Du, R.; Li, W.; Chen, S.; Li, C.; Zhang, Y. Unpaired Underwater Image Enhancement Based on CycleGAN. *Information* **2022**, *13*, 1. [CrossRef]

23. Dubreuil, M.; Delrot, P.; Leonard, I.; Alfalou, A.; Brosseau, C.; Dogariu, A. Exploring underwater target detection by imaging polarimetry and correlation techniques. *Appl. Opt.* **2013**, *52*, 997–1005. [CrossRef] [PubMed]
24. Zhu, J.Y.; Park, T.; Isola, P.; Efros, A.A. Unpaired Image-to-Image Translation Using Cycle-Consistent Adversarial Networks. In Proceedings of the 2017 IEEE International Conference on Computer Vision (ICCV), Venice, Italy, 22–29 October 2017; pp. 2242–2251.

photonics

MDPI

Article

An Adaptive Denoising Method for Photon-Counting LiDAR Point Clouds: Application in Intertidal Zones

Cheng Wu [1,2], Lei Ding [3], Lin Cong [1] and Shaoning Li [2,*]

1 Key Laboratory of SmartEarth, Beijing 100029, China; wc@mail.hnust.edu.cn (C.W.)
2 National-Local Joint Engineering Laboratory of Geo-Spatial Information Technology, Hunan University of Science and Technology, Xiangtan 411201, China
3 Space Star Technology Co., Ltd., Xi'an 710016, China
* Correspondence: lsn@hnust.edu.cn; Tel.: +86-073158290092

Abstract: The intertidal zone, as a dynamic ecosystem at the interface of land and sea, plays a critical role in environmental protection and disaster mitigation. The Ice, Cloud, and Land Elevation Satellite-2 (ICESat-2) is equipped with the Advanced Topographic Laser Altimeter System (ATLAS) with the ability to penetrate the water bodies, enabling its use for bathymetric measurements. However, the complex land cover types and frequent environmental changes in intertidal zones pose significant challenges for precise measurement and dynamic monitoring. In an effort to address the denoising challenges of ICESat-2 photon point cloud data in such complex environments, this study proposes an adaptive photon denoising method that is capable of dynamically adjusting the denoising strategy for different types of photon data. ATL03 data from four typical intertidal zones were selected for denoising experiments. The results indicated that the proposed adaptive denoising method achieved average recall, precision, and F-score values of 0.9885, 0.9927, and 0.9906, respectively, demonstrating excellent denoising performance and stability. This method provides an effective data processing approach for high-precision monitoring of intertidal zone topography.

Keywords: photon-counting LiDAR altimetry; intertidal zone; adaptive denoising; ICESat-2; accuracy analysis

1. Introduction

The intertidal zone, located between the mean low tide line and the mean high tide line of spring tides, undergoes periodic submersion and exposure due to tidal actions. This dynamic environment exhibits complex erosion and deposition patterns, including the formation of tidal creeks by water flow erosion and pits by wave erosion. As a critical region for ecological protection and disaster mitigation, the intertidal zone holds significant ecological and environmental value [1–3]. Therefore, it is of great significance to dynamically monitor the intertidal zones and timely detect their topographic changes.

Traditional single-beam or multi-beam bathymetric instruments rely on shipboard platforms. Although the operation is straightforward, they require extensive human and material resources and are susceptible to grounding in shallow waters, resulting in low measurement efficiency. Such limitations hinder their effectiveness in achieving continuous, large-scale observations in the dynamic and complex intertidal zones [4]. Airborne bathymetric light detection and ranging (LiDAR) systems provide a balance between accuracy and efficiency, but the cost of data acquisition is high and often restricted by airspace regulations. These methods are limited in remote or sensitive areas where

https://doi.org/10.3390/photonics12010013

ships and aircraft find it difficult to approach [4,5]. With the development of spaceborne remote sensing technology, various remote sensing methods have been used to obtain underwater terrain information. Optical remote sensing provides abundant data sources with wide coverage, which can periodically and widely acquire optical information about waters, such as hyperspectral and multispectral remote sensing images [6,7]. However, the measurement accuracy is relatively low due to various factors such as imaging technology and weather conditions [8]. Synthetic Aperture Radar (SAR) signals can penetrate clouds but are unable to penetrate the water bodies vertically; it can only obtain information about the underwater bottom based on surface wave patterns, restricting its bathymetric accuracy [5].In, contrast, spaceborne LiDAR offers advantages such as wide coverage, high repeat observation frequency, minimal impact from external factors, and high measurement accuracy. These advantages have made it widely applied in remote sensing studies of lakes, glaciers, and oceans [9]. ICESat-2, launched by NASA in September 2018, is a new-generation spaceborne LiDAR satellite equipped with the ATLAS sensor. ATLAS utilizes multi-beam, micropulse, and photon-counting technologies, featuring a higher pulse repetition frequency to obtain small-footprint, high-density photon point cloud data. Its maximum bathymetric capability can reach 38 m, showing great potential in the ecological environmental monitoring of intertidal zones [10,11]. However, the ATLAS system's weak laser signals are susceptible to noise from the solar background, system characteristics, atmospheric scattering, and ground cover. These factors lead to a large amount of noise in the received data, complicating the signal extraction process [12]

As a region of frequent interaction between seawater and land, the intertidal zone contains numerous suspended particles and organisms. Photon data usually include signals reflected from various land covers such as land, water bodies, and vegetation, increasing the difficulty of noise removal. The Density-Based Spatial Clustering of Applications with Noise (DBSCAN) extracts signal photons based on the distribution density differences between signal photons and noise photons and has been widely applied in denoising various types of point cloud data [13]. However, this algorithm is highly sensitive to parameter settings, making it difficult to adapt to the complex environmental characteristics of the intertidal zone, thereby affecting denoising performance [14]. Denoising algorithms based on raster image processing require point cloud data to be rasterized, which may result in significant information loss in areas with complex land cover types, such as intertidal zones, and consequently lead to a notable reduction in denoising accuracy [15]. The statistical histogram algorithm identifies and removes noise points by calculating the local distance between each point and its neighboring points, plotting a local distance histogram, and setting thresholds. It is mainly applied in forests and mountainous areas [16]. However, in intertidal zones, the distribution characteristics of signal photons and noise photons are not as regular as those in forests or mountainous areas, and photon distribution often exhibits higher dynamics [17]. This limits the denoising capability of the statistical histogram algorithm. The ATL08 product employs the Differential, Regressive, and Gaussian Adaptive Nearest Neighbor (DRAGANN) method to denoise ATL03 data and performs well in relatively stable terrestrial areas with vegetation and minimal terrain variation [18]. However, when applied to the intertidal zone with its complex water–land transition topography and highly dynamic environment, the denoising performance of this method may decrease. Although these denoising methods have achieved some results, most of them are for a single scene and are mainly applied to land areas. When applied to complex and variable environments, they may have limitations. Research on denoising in intertidal zones with their complex and diverse land cover types is relatively scarce, which limits the capability to accurately monitor intertidal zones. Therefore, effectively removing noise in intertidal zones to obtain accurate land cover and topographic information has become a critical issue that needs to be solved urgently.

To improve the accuracy of photon denoising in intertidal zones and retain more effective land cover and topographic information, this study proposes a photon denoising method based on an adaptive strategy. By analyzing the photon distribution characteristics at the water–land boundary in intertidal zones, the method performs preliminary classification and adaptively adjusts the denoising strategy according to different land cover types, selecting the most suitable denoising algorithm for different types of photon data. The proposed method is further compared with traditional denoising algorithms, including DBSCAN, the statistical histogram algorithm, and the DRAGANN algorithm, to evaluate its denoising accuracy.

2. Study Areas and Data

ICESat-2/ATLAS operates with a 532-nanometer laser to emit pulses at a frequency of 10 kHz. Considering the satellite's orbital altitude and velocity, the along-track photon spacing is approximately 0.7 m, forming an almost continuous profile on the Earth's surface. As illustrated in Figure 1, each laser pulse emitted by the laser is split by optical elements into six individual pulses, arranged parallel on the ground into three pairs. Each pair of adjacent beams has a cross-track spacing of about 3.3 km. Within each pair, there are two sub-beams—namely, a strong beam and a weak beam—with an energy ratio of approximately 4:1. The strong and weak beams have a cross-track spacing of about 90 m and an along-track spacing of about 2.5 km, enhancing spatial coverage and determining local across-track slopes [19,20]. The main parameters of ICESat-2/ATLAS are given in Table 1. ICESat-2 provides four different levels of data products (L1, L2, L3A, and L3B) free to the public. The ATL03 product provides the source data for intertidal zone denoising in this study; it is a Level 2 product that records the time, latitude, longitude, and elevation of each photon event detected by ATLAS [21].

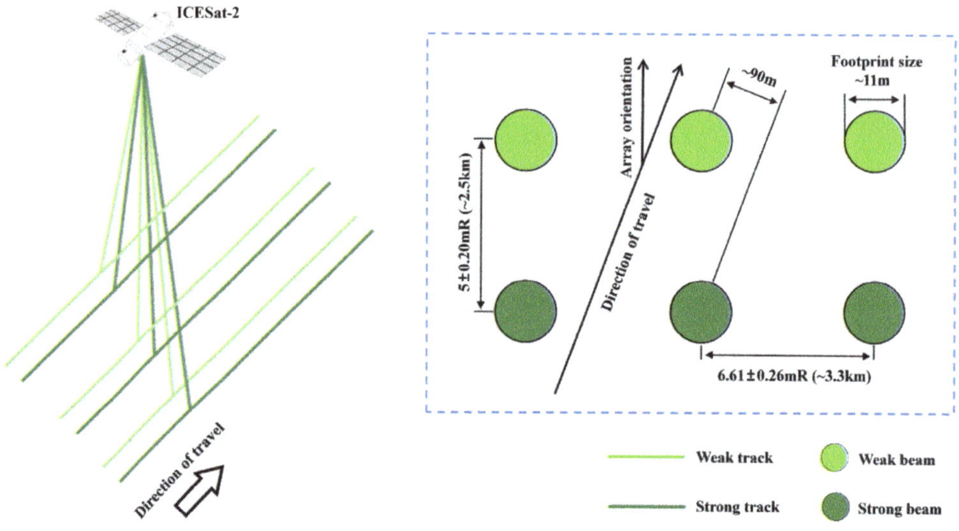

Figure 1. ICESat-2 tracks and footprint distribution.

Table 1. Main parameters of ICESat-2/ATLAS.

Parameters	Values	Parameters	Values
Track repeat period	91 days	Along-track separation	0.7 m
Laser wavelength	532 nm	Number of beams	6
Transmitted pulse width	1.5 ns	Distance with a pair	~90 m
Pulse repetition rate	10 kHz	Distance between beam pairs	~3.3 km
Footprint size	~11 m	Beam energy per pulse (strong, weak)	175 ± 17 μJ, 45 ± 5 μJ

The ATL03 data used in this study are available from the National Snow and Ice Data Center (NSIDC) website (https://nsidc.org/data/icesat-2/data, accessed on 20 April 2024). Considering the impact of land cover and topographic factors on denoising effectiveness, four typical intertidal zones were selected for experiments in this study. The sandy intertidal zone is characterized by expansive sandy beaches formed by the accumulation of fine sand. The terrain is usually flat with relatively simple biological coverage (see Figure 2A). The rocky intertidal zone is typically found along coastlines composed of rocks. Due to its hard geological characteristics, this type of intertidal zone is less affected by erosion (see Figure 2B). The coral reef intertidal zone, as shown in Figure 2C, is submerged during high tide and exposed during low tide. The estuarine intertidal zone is commonly found in river estuaries (see Figure 2D). Sediments brought by rivers accumulate to form rich sandy or muddy terrains, often exhibiting high biodiversity.

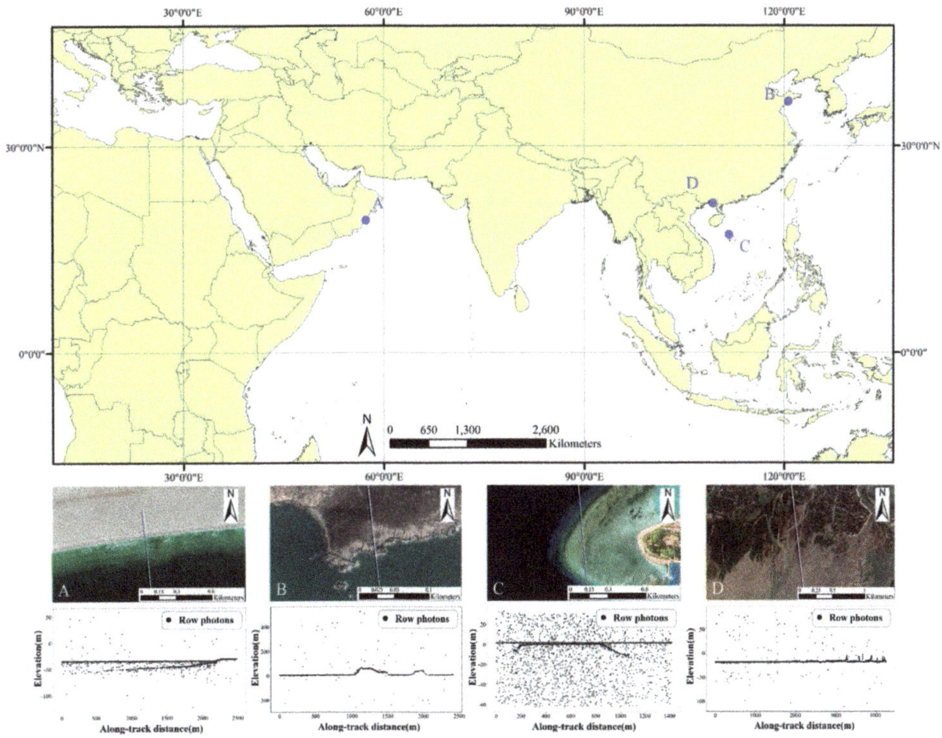

Figure 2. Laser tracks and profiles of the study areas: (**A**) sandy intertidal zone; (**B**) rocky intertidal zone; (**C**) coral reef intertidal zone; (**D**) estuarine intertidal zone, with the solid blue line as the laser track.

Detailed information on the four ATL03 data tracks used in this study is presented in Table 2, including the corresponding intertidal zone types, study areas, acquisition times, and the pulse beams used.

Table 2. Study areas and data.

Intertidal Zone Types	Study Areas	ATL03 Data	Day/Night	Track Used
Sandy intertidal zone	Hayma, Oman	ATL03_20231216172359_13552101_006_02_h5	night	gt1r
Rocky intertidal zone	Qingdao, Shandong	ATL03_20220917105543_13371602_006_01.h5	night	gt2r
Coral reef intertidal zone	Sansha, Hainan	ATL03_20210717075633_03621201_006_01.h5	day	gt1r
Estuarine intertidal zone	Guangxi Beihai	ATL03_20190929152507_00420501_006_02.h5	night	gt1r

3. Methods

The intertidal zones contain various land covers such as land, water bodies, and vegetation. The interaction of laser emission signals with seawater and land results produces complex reflected signals, increasing the difficulty of distinguishing between signal and noise in photon point clouds [22]. To solve this problem, we first carried out a preliminary classification based on the distribution characteristics of photons at the water–land boundary. Subsequently, denoising algorithms were adaptively selected and applied according to the unique characteristics of different types of photons.

3.1. Preliminary Classification

In intertidal zones, the density of water surface photons is usually higher than that of exposed land and vegetated areas. This is mainly because water surface photon signals are obtained from the combined reflections of the water surface and the underwater, resulting in a larger number of water surface photons that are densely distributed [23]. In land areas close to the water surface, the reflective properties of the land surface are easily affected by factors such as vegetation coverage and surface slope, making land photons relatively sparse. Therefore, based on the non-uniformity of photon density distribution, we carried out a preliminary classification of photons at the water–land boundary. Considering the limited bathymetric capability of ICESat-2 when measuring intertidal zones, the boundary line near the shore can be approximated as a horizontal line [24]. The horizontal dividing line is calculated as follows:

$$d = lm - \Delta \tag{1}$$

where lm and σ are the mean elevation and standard deviation of the local water surface photons. Δ was set to 3σ to ensure sufficient distinction between water surface photons and underwater photons. As illustrated in Figure 3, photons above the dividing line were classified as water surface photons, while those below were classified as underwater photons.

At the water–land boundary, the photon density of the water surface is usually higher than that of the land. The moving window statistical method was used, setting the width of the moving window to 5 m and the length to be the difference in elevation between the photon with the highest elevation and the photon with the lowest elevation of the photon data in this track. We searched along the photon track and counted the histogram of the photon elevation frequency distribution within each window, as shown in Figure 4.

By analyzing the photon point cloud data in Figure 4, it was found that water surface areas (the yellow rectangular box) usually exhibit a large number of photons and a concentrated elevation distribution. In contrast, land areas (the red rectangular box) have relatively fewer photons with more dispersed elevation distribution. Based on this phenomenon, the concentration of the photon elevation distribution was used as a key parameter to distinguish the water surface from the land areas. This method takes full use of the relatively flat surface of the water while the land is characterized by a large number of features and obvious changes in topography. Specifically, the concentration of photon elevation distribution is calculated as follows:

$$f = 1/(1+s) \quad (2)$$

where s is the standard deviation of photons detected within the moving window. The function f normalizes the result to the range of (0,1], making comparisons between different regions more intuitive and consistent. An f value close to 1 indicates a highly concentrated photon elevation distribution, corresponding to water surface areas; an f value close to 0 indicates a highly dispersed distribution, corresponding to land areas. By setting a threshold for f, we can distinguish between water surface photons and land photons.

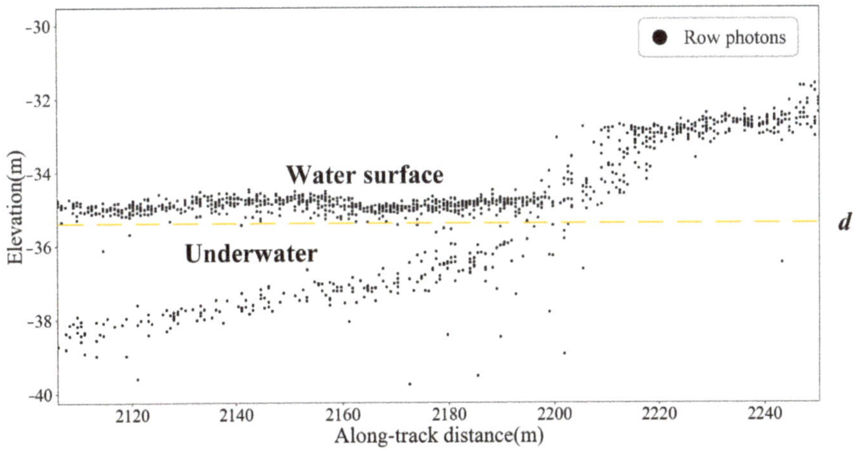

Figure 3. Schematic of ATL03 raw data boundary.

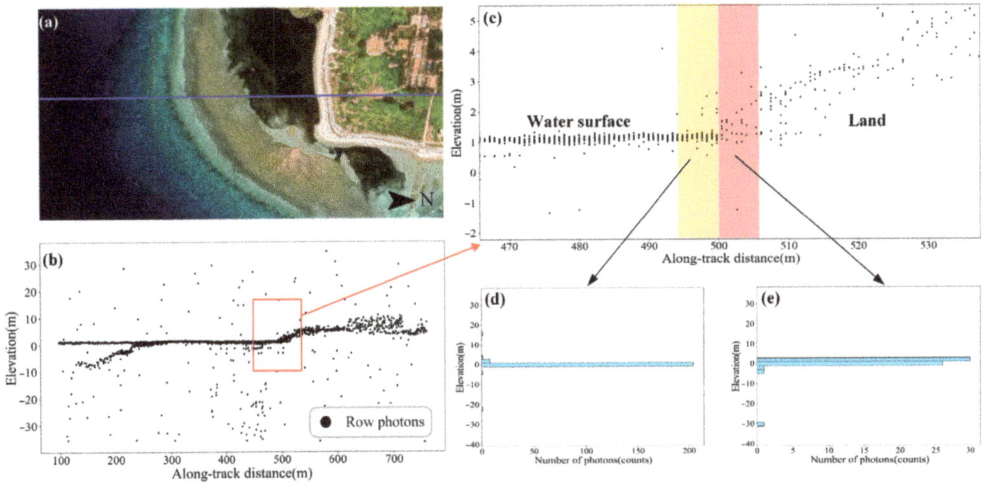

Figure 4. (**a**) Solid blue line is the laser track. (**b**) ATL03 raw data profile with moving window in red. (**c**) Schematic of the moving window search. (**d**) Histogram of the frequency distribution of photon elevation within the yellow rectangular box. (**e**) Histogram of photon elevation frequency distribution in the red rectangular box.

We carried out a preliminary classification of the four types of intertidal zones: The sandy intertidal zone, as shown in Figure 5a, includes water surface photons, underwater photons, and land photons. The rocky intertidal zone, as shown in Figure 5b, includes water surface photons and land photons. The coral reef intertidal zone, as shown in Figure 5c, includes water surface photons and underwater photons. The estuarine intertidal zone is characterized by a substantial presence of suspended particles or organisms, which results in non-target reflective noise blending with photon signals from both the water surface and the underwater bottom. This phenomenon increases the photon thickness and leads to a mixture of water surface and underwater terrain photon signals. The results of the preliminary classification are presented in Figure 5d, which includes water surface photons and land photons. Each category of photons in all these intertidal zones contains both signal photons and noise photons.

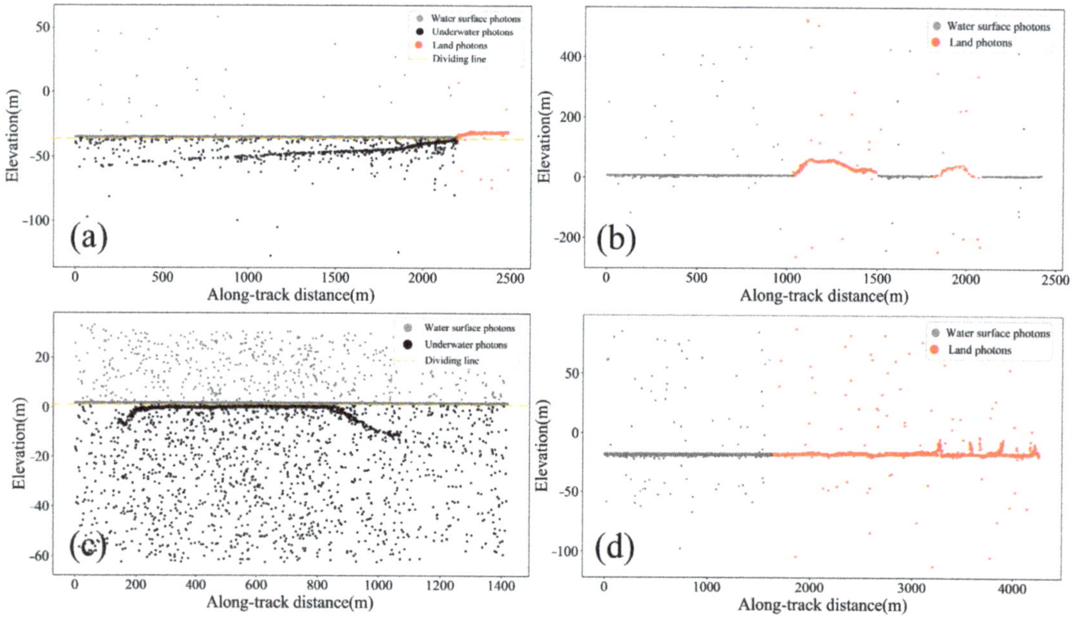

Figure 5. Preliminary classification results of photon data: (**a**) sandy intertidal zone; (**b**) rocky intertidal zone; (**c**) coral reef intertidal zone; (**d**) estuarine intertidal zone.

3.2. Adaptive Denoising

In land areas, factors such as vegetation coverage and surface slope result in a relatively sparse density of land photons. Therefore, a histogram distance-weighted statistical algorithm was utilized for denoising land photons. In contrast, in water surface regions, the photon density is higher, exhibiting pronounced variations in density in the along-track and cross-track directions. In underwater terrain regions, due to the long-term erosion by seawater, the extension of underwater terrain is relatively smooth, with fewer fluctuations compared with land. The underwater terrain photons have similar distribution characteristics to the water surface photons [25]. Therefore, an adaptive elliptical DBSCAN algorithm was used to effectively denoise water surface and underwater photons.

After one round of denoising the raw photon data, evaluation metrics, namely surface continuity (*SC*) and coefficient of variation (*CV*) for neighborhood distances, were introduced to establish an iterative feedback mechanism for assessing denoising results.

This approach contributes to the continuous optimization of denoising outcomes through iterative parameter adjustment.

Under natural conditions, the variation in the elevation of water surface photons tends to be smooth and stable, and the variation in the elevation of underwater terrain photons is relatively gradual. To evaluate the elevation continuity of both water surface and underwater terrain photons, and to avoid signal fluctuations caused by residual noise, the standard deviation of photon elevation can be used as an indicator. The surface continuity (SC) is calculated as follows:

$$SC = \sqrt{\frac{1}{N}\sum_{i=1}^{N}\left(h_i - \overline{h}\right)^2} \tag{3}$$

where h_i represents the elevation of the i-th photon, $\overline{h}$ is the mean elevation of all photons, and N represents the total number of photons.

In contrast, the variations in the elevation of land photons are more pronounced, yet photons of the same type exhibit spatial consistency, with distances between neighboring photons oscillating within a certain range. To evaluate the spatial distribution consistency of photon data, the coefficient of variation (CV) for neighborhood distances is calculated as follows:

$$CV = \frac{\sigma_{dist}}{\mu_{dist}} \tag{4}$$

where σ_{dist} is the standard deviation of distances between neighboring photons, and μ_{dist} is the mean distance between neighboring photons.

To define appropriate threshold ranges, 10–20 tracks of manually denoised photon point cloud data outside the study areas were selected to calculate the mean and standard deviation of SC and CV values, as shown below.

$$\overline{SC} = \frac{1}{n}\sum_{i=1}^{n} SC_i,\ \overline{CV} = \frac{1}{n}\sum_{i=1}^{n} CV_i \tag{5}$$

$$\sigma_{SC} = \sqrt{\frac{1}{n}\sum_{i=1}^{n}\left(SC_i - \overline{SC}\right)^2},\ \sigma_{CV} = \sqrt{\frac{1}{n}\sum_{i=1}^{n}\left(CV_i - \overline{CV}\right)^2} \tag{6}$$

where SC_i and CV_i represent the SC and CV values of the i-th track, and n is the number of point cloud tracks.

Based on the calculated mean and standard deviation, threshold ranges are defined as follows:

$$SC_{threshold} = \overline{SC} + 3\sigma_{SC},\ CV_{threshold} = \overline{CV} + 3\sigma_{CV} \tag{7}$$

To more rigorously retain signal photons, three times the standard deviations σ_{SC} and σ_{CV} were selected.

After denoising the dataset, threshold evaluation was applied to determine whether to retain the photon as a signal photon. The criterion was as follows:

$$\begin{cases} SC \le SC_{threshold} \\ CV \le CV_{threshold} \end{cases} \tag{8}$$

Here, if the SC and CV values satisfy the threshold conditions, the denoised data are considered to have met the retention criteria for signal photons, and the iteration stops, outputting the data as signal photons. To avoid infinite loops, the maximum number of denoising iterations was set to 3; that is, if the SC and CV values still did not reach the threshold after three iterations, the photon was considered a signal photon and output. The basic flowchart of the adaptive denoising method is shown in Figure 6.

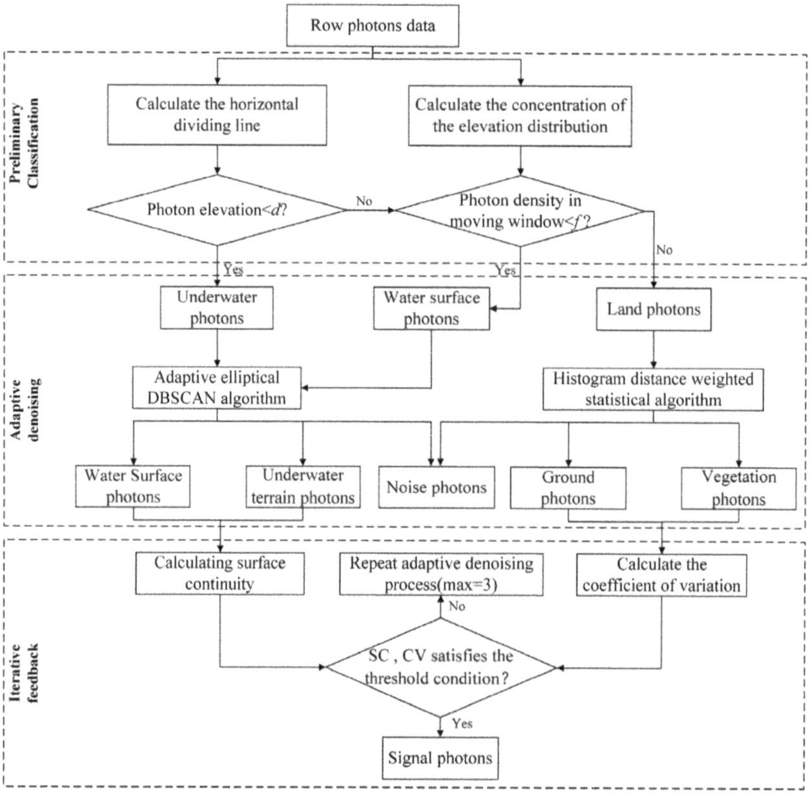

Figure 6. Basic flowchart of adaptive denoising method.

The histogram distance-weighted statistical algorithm distinguishes noise points from non-noise points in point cloud data based on differences in spatial distribution. Its advantage lies in its precise processing capability for low-density point cloud data. By dynamically adjusting weights and distance calculations, it can effectively distinguish between noise and signal photons, reducing the error recognition rate [26].

The Euclidean distance $dist_i$ between each point i in the point cloud and its K nearest neighboring points j is calculated as follows:

$$dist_i = \sqrt{(x_i - x_j)^2 + (y_i - y_j)^2} \tag{9}$$

where x and y are the along-track distance and elevation of the photon point cloud.

To more accurately identify noise points, these distances were assigned weights [27]. The weight function is shown in the following equation:

$$weight = 1 - \exp\left(\frac{-dist_i^{\,2}}{\sigma^2}\right) \tag{10}$$

$$\sigma^2 = \frac{\sum_{j=1}^{i=1} (x_i - x_j)^2 + (y_i - y_j)^2}{K} \tag{11}$$

where σ^2 is used to adjust the attenuation rate of the weight, ensuring that the larger the distance between two points, the larger the weight; conversely, the smaller the distance, the smaller the weight.

The total weighted distance D_i of each point to its K nearest neighbors is calculated as follows:

$$D_i = \sum_{j=1}^{i=1} \sqrt{(x_i - x_j)^2 + (y_i - y_j)^2} \cdot weight \tag{12}$$

Based on these total distances, a cumulative weighted distance frequency distribution histogram was plotted. Due to the different densities of noise points and non-noise points, the histogram exhibits two peaks, allowing for threshold setting to remove noise points.

The adaptive elliptical DBSCAN algorithm improves upon the traditional DBSCAN algorithm [28]. Its core idea is to replace the circular neighborhood in DBSCAN with an elliptical neighborhood, adaptively adjusting clustering parameters based on the local density characteristics of the data. This enhancement enables more effective handling of data with significant differences in distribution characteristics in the horizontal and vertical directions and uneven density distribution. In the traditional DBSCAN algorithm, two parameters need to be predefined: the neighborhood radius (ε) and the minimum number of core points ($MinPts$). The adaptive elliptical DBSCAN algorithm adaptively adjusts the neighborhood radius (ε) and $MinPts$ to better accommodate variations in local photon density. When traversing the dataset, each photon is treated as a potential clustering object. If the number of photons within the given neighborhood radius ε is not less than the threshold $MinPts$, the photon is considered a core point. Subsequently, these core points and the photons in their neighborhoods are clustered based on density connectivity.

The elliptical distance between any two points $p(x_p, y_p)$ and $q(x_q, y_q)$ in two-dimensional space is calculated as follows:

$$dist(p,q) = \sqrt{\left(\frac{x_p - x}{a}\right)^2 + \left(\frac{y_p - y_p}{b}\right)^2} \tag{13}$$

When $dist(p,q) \leq \varepsilon$, point q is within the neighborhood $N_\varepsilon(p)$ [11]. To simplify parameters, the neighborhood radius was fixed at $\varepsilon = 1$. The size of the elliptical neighborhood was determined by the semi-major axis a and the semi-minor axis b.

Using the statistical characteristics of local photon data, the adaptive values of a and b are calculated as follows:

$$a = \varepsilon \times \delta x,\ b = \varepsilon \times \delta y \tag{14}$$

where δx and δy are the standard deviation of photons in the data segment in the horizontal and vertical directions.

The average photon density ρ within the elliptical neighborhood is calculated as follows:

$$\rho = \frac{\pi abM}{HL} \tag{15}$$

where M is the total number of photons in the data segment, H is the elevation range within the segment, and L is the length along the track direction.

The adaptive $MinPts$ is calculated as follows:

$$MinPts = \gamma \cdot \rho \tag{16}$$

where $\gamma \geq 1$, and the value of γ is estimated empirically so as to adaptively adjust $MinPts$.

For the photon point cloud data in the study areas, as long as the photon density in its neighborhood exceeds $MinPts$, these photons were classified as signal photons; otherwise, they were marked as noise photons. Figure 7 illustrates the denoising process of this adaptive elliptical BDSCAN algorithm.

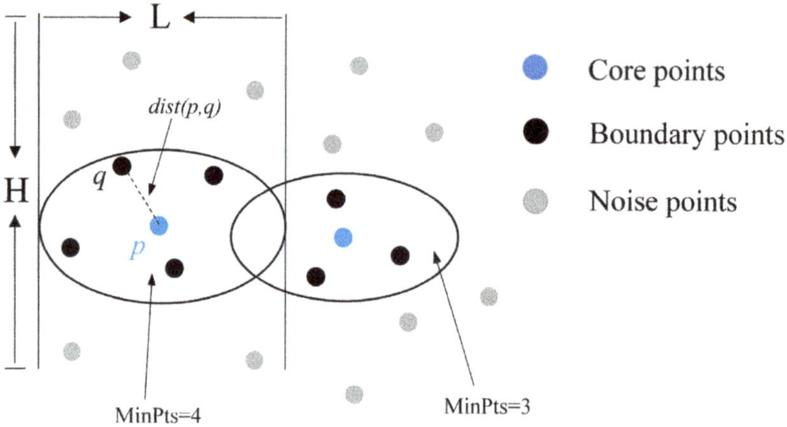

Figure 7. Schematic diagram of the adaptive elliptical DBSCAN algorithm.

4. Discussion

4.1. Analysis of Denoising Results

The proposed adaptive denoising method was applied to various intertidal zones, and the results are shown in Figure 8. This method effectively removes noise while preserving the integrity of signal photons. In estuarine intertidal zones, abundant sediment and vegetation are often present. By referencing classification results from ATL08 data, the denoised land photons were further classified into vegetation and ground photons.

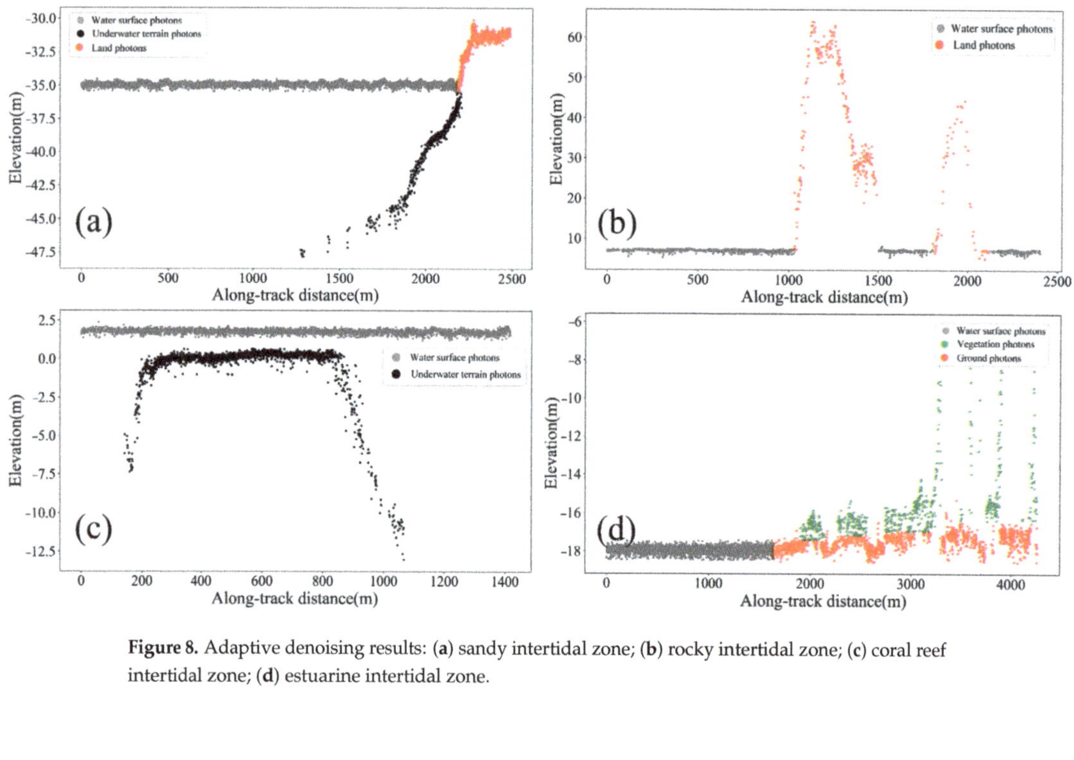

Figure 8. Adaptive denoising results: (**a**) sandy intertidal zone; (**b**) rocky intertidal zone; (**c**) coral reef intertidal zone; (**d**) estuarine intertidal zone.

The adaptive denoising method proposed in this paper incorporates improvements to the DBSCAN and statistical histogram algorithms, and the comparison with these two algorithms can intuitively reflect the effect of the improvements; the DRAGANN algorithm, as the official algorithm for ICESat-2 data processing, is highly representative. Therefore, the denoising results of the adaptive method were compared with those of traditional DBSCAN, statistical histogram, and DRAGANN algorithm across different types of intertidal zones. The DBSCAN algorithm, widely used for spatial clustering analysis, is particularly effective in handling noisy datasets. In the original ATL03 photon data, signal photons exhibit higher spatial density compared to noise photons. The DBSCAN algorithm uses this characteristic to cluster signal photons while identifying and removing dispersed noise photons [29]. The statistical histogram method leverages the differences in spatial distribution between signal photons and noise photons. It distinguishes signal photons from noise photons by analyzing spatial distribution differences, calculating photon distances, performing frequency analysis, and generating histograms [30]. The DRAGANN algorithm is a denoising algorithm specifically designed for the ATL08 product. It distinguishes signal photons and noise photons through adaptive nearest neighbor search and bimodal distribution analysis [14].

The denoising results of the sandy intertidal zone are shown in Figure 9. The proposed adaptive denoising method effectively removes most of the noise and better retains the underwater terrain signals. At the water–land boundary shown in Figure 9e, the adaptive denoising method has essentially removed all noise; however, with the DBSCAN algorithm, there are still scattered noises around the signal photons. The statistical histogram algorithm still retains more noise photons. The DRAGANN algorithm demonstrates good denoising performance, but may mistakenly identify underwater terrain signal photons as noise and remove them. In the red box area, the adaptive denoising method can more completely recognize underwater terrain photons; the DBSCAN algorithm identifies part of the underwater terrain photons as noise and removes them. Although the statistical histogram method can fully retain underwater terrain photons, it also retains more noise. The DRAGANN algorithm exhibits over-denoising, resulting in most of the underwater terrain signal photons being incorrectly removed.

The denoising results of the rocky intertidal zone are shown in Figure 10. When denoising water surface photons, both the proposed adaptive denoising method and the DRAGANN algorithm can effectively eliminate the vast majority of noise, while the DBSCAN algorithm and the statistical histogram algorithm retain more underwater noise (see Figure 10e). In the red box area, the adaptive denoising method, DBSCAN algorithm, and statistical histogram algorithm can all better retain land signal photons, whereas the DRAGANN algorithm has over-denoising issues, leading to most of the surface water and land signal photons being incorrectly removed. In contrast, the proposed adaptive denoising method not only effectively eliminates most of the noise but also better preserves land signal photons.

The denoising results of the coral reef intertidal zone are shown in Figure 11. A detailed analysis of the denoising performance of the four methods indicates that both the DBSCAN and DRAGANN algorithms effectively remove the majority of noise; however, a small amount of residual noise persists between the water surface and underwater terrain (Figure 11e). The statistical histogram algorithm retains a higher level of noise compared to the other methods, whereas the proposed adaptive denoising method demonstrates superior performance in eliminating noise photons in this region. In the red box area, although the DBSCAN algorithm can effectively eliminate noise, a small portion of signal photons are mistakenly identified as noise and removed. Similarly, the statistical histogram algorithm, though better at retaining underwater terrain signal photons, exhibits weaker denoising performance in areas where the underwater terrain is close to the water surface. The DRAGANN algorithm is less effective in retaining underwater terrain photons, resulting in a loss of critical signal data. In contrast, the adaptive denoising algorithm can accurately remove surrounding noise while completely preserving underwater terrain signal photons.

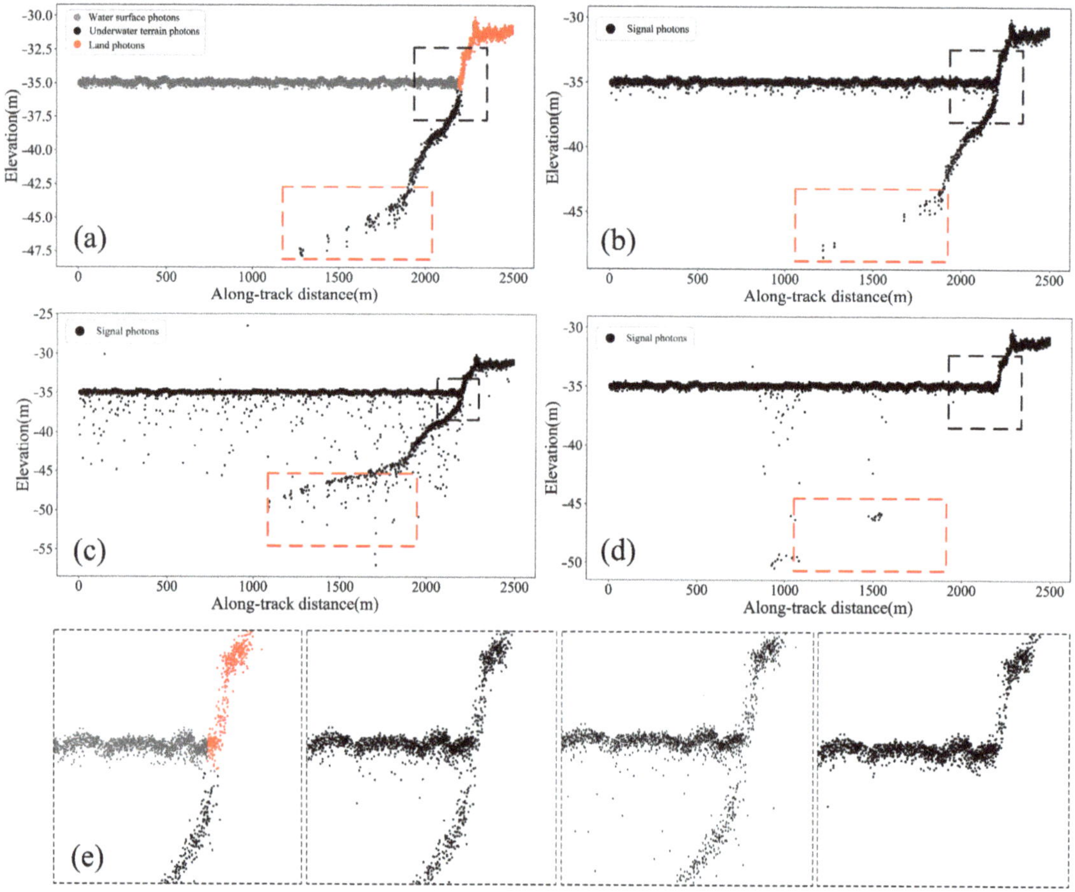

Figure 9. Sandy intertidal zone denoising results: (**a–d**) the adaptive denoising method, DBSCAN algorithm, histogram statistical algorithm, and DRAGANN algorithm in order; (**e**) zoomed-in comparison of the black dashed box, from left to right, the adaptive method, DBSCAN algorithm, histogram statistic algorithm, and DRAGANN algorithm.

The denoising results of the estuarine intertidal zone are shown in Figure 12. In this environment, the mixing of water surface photons with underwater terrain signal photons results in increased signal photon thickness. The proposed adaptive denoising method, as well as the DBSCAN and DRAGANN algorithms, effectively removes a significant portion of apparent noise, while the statistical histogram algorithm retains some scattered noise. In the red box area, both the adaptive denoising method and the DRAGANN algorithm demonstrate superior performance in noise removal. In contrast, the DBSCAN and statistical histogram algorithms retain more underwater noise, reducing the overall effectiveness. At the water–land boundary in Figure 12e, the DBSCAN algorithm and the statistical histogram algorithm retain more noise, while the adaptive denoising method and the DRAGANN algorithm can better remove the noise around the signal photons.

In summary, across the four different types of intertidal zones, the proposed adaptive denoising method can comprehensively remove noise, clearly distinguish between signal and noise, and maintain the integrity of signal photons.

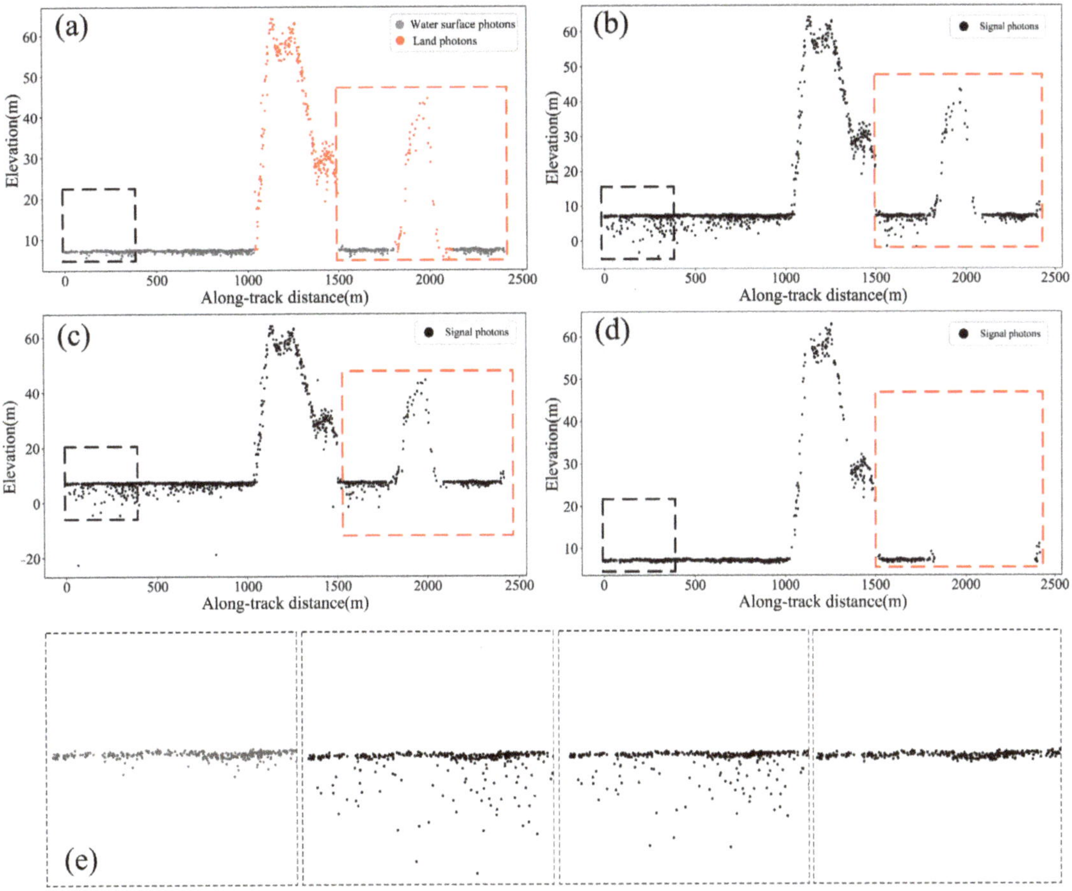

Figure 10. Rocky intertidal zone denoising results: (**a**–**d**) the adaptive denoising method, DBSCAN algorithm, histogram statistical algorithm, and DRAGANN algorithm in order; (**e**) zoomed-in comparison of the black dashed box, from left to right, the adaptive method, DBSCAN algorithm, histogram statistic algorithm, and DRAGANN algorithm.

4.2. Accuracy Evaluation

This paper introduces three statistical evaluation metrics to quantitatively assess the performance of the algorithms [31]: recall (R), precision (P), and F-score (F). Recall represents the proportion of correctly extracted effective signal photons to the total number of original signal photons. Precision refers to the ratio of the number of correctly extracted effective signal photons to the total number of extracted effective signal photons. The F-score is the harmonic mean of recall and precision. The calculation formulas are shown in the following equations:

$$R = \frac{TP}{TP + FN} \tag{17}$$

$$P = \frac{TP}{TP + FP} \quad (18)$$

$$F = \frac{2PR}{P + R} \quad (19)$$

where TP is the number of photons identified as signal photons, which are actually signal photons; FP is the number of photons identified as signal photons, but they are actually noise photons; FN is the number of photons identified as noise photons, but they are actually signal photons.

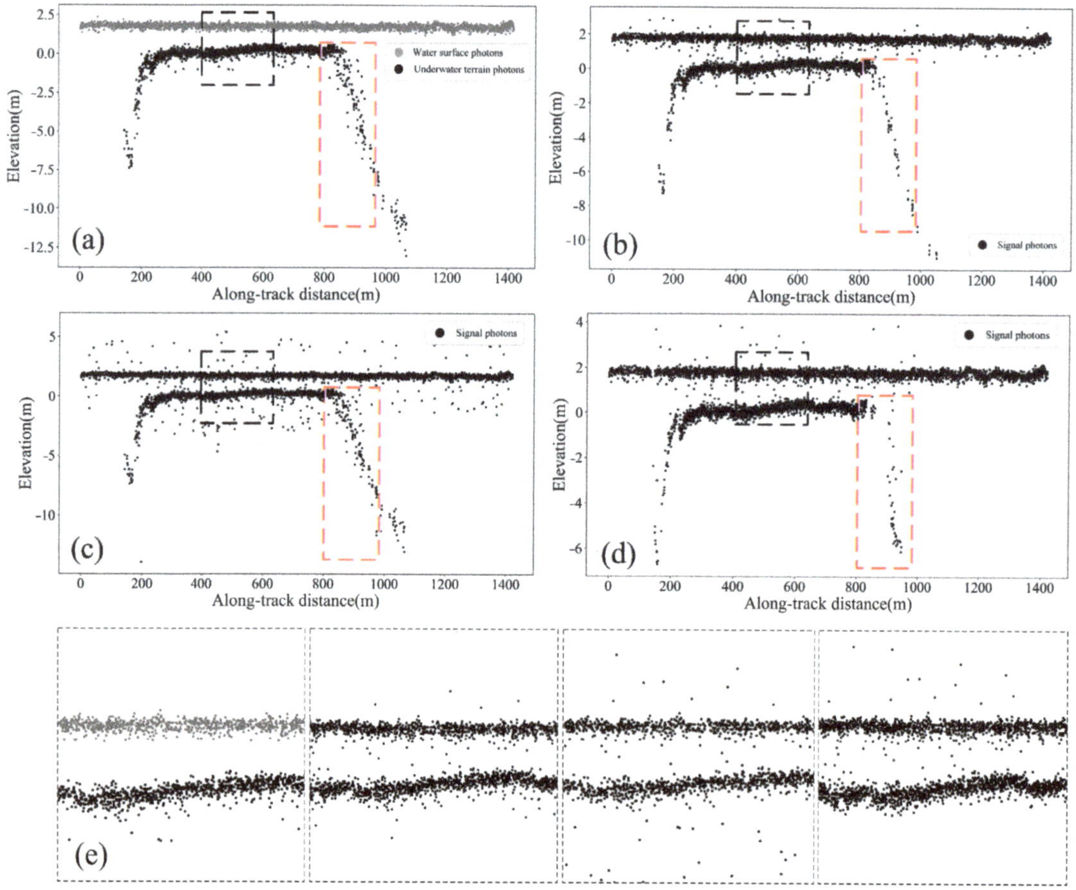

Figure 11. Coral reef intertidal zone denoising results: (**a**–**d**) the adaptive denoising method, DBSCAN algorithm, histogram statistical algorithm, and DRAGANN algorithm in order; (**e**) zoomed-in comparison of the black dashed box, from left to right, the adaptive method, DBSCAN algorithm, histogram statistic algorithm, and DRAGANN algorithm.

In this study, photon data from four typical intertidal zones—sandy, rocky, coral reef, and estuarine intertidal zones—were subjected to denoising processing using the adaptive denoising method, the DBSCAN algorithm, the statistical histogram algorithm, and the DRAGANN algorithm. The denoising effects were evaluated and statistically analyzed using the three metrics R, P, and F, with the results shown in Table 3 and Figure 13.

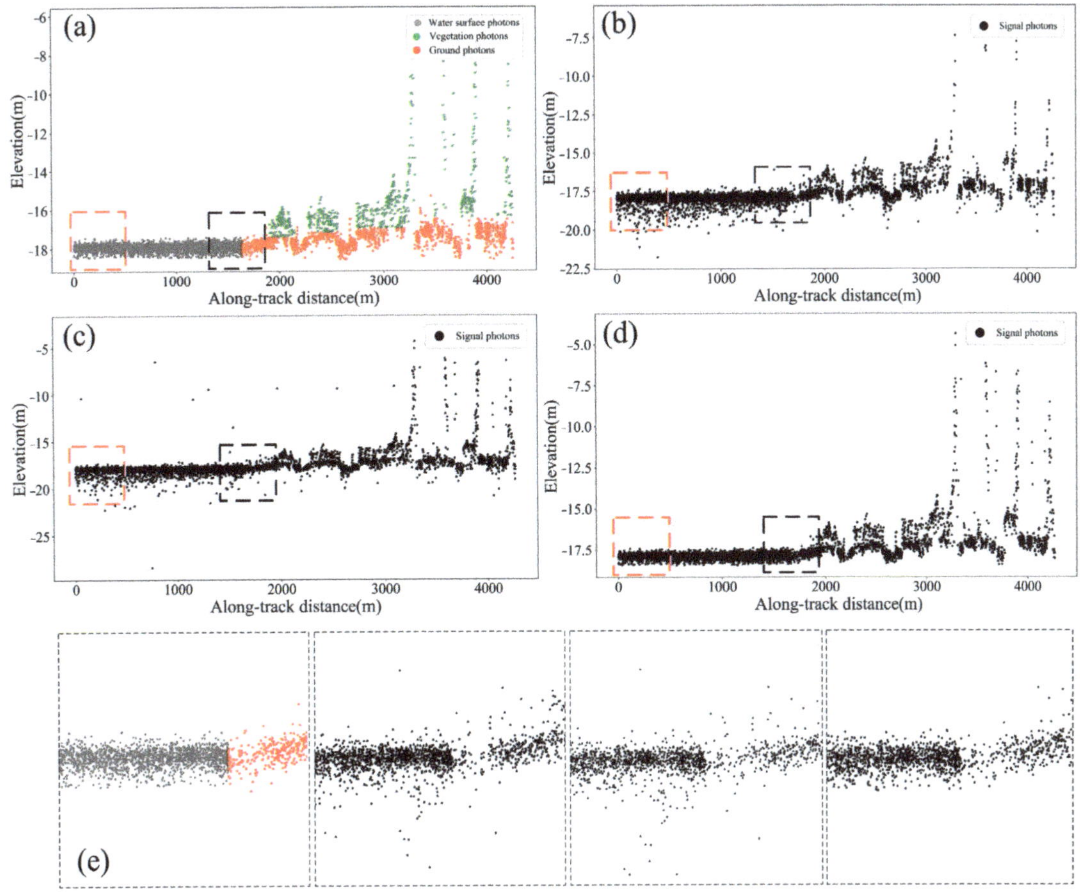

Figure 12. Estuarine intertidal zone denoising results: (**a**–**d**) the adaptive denoising method, DBSCAN algorithm, histogram statistical algorithm, and DRAGANN algorithm in order; (**e**) zoomed-in comparison of the black dashed box, from left to right, the adaptive method, DBSCAN algorithm, histogram statistic algorithm, and DRAGANN algorithm.

Table 3. Statistical table of denoising metrics.

Study Areas	Denoising Method	*R*	*P*	*F*
Sandy intertidal zone	Adaptive	0.9853	0.9930	0.9891
	DBSCAN	0.9625	0.9853	0.9738
	Statistical histogram	0.9878	0.9522	0.9697
	DRAGANN	0.9141	0.9888	0.9500
Rocky intertidal zone	Adaptive	0.9818	0.9848	0.9833
	DBSCAN	0.9968	0.8716	0.9300
	Statistical histogram	0.9984	0.8953	0.9440
	DRAGANN	0.7415	0.9870	0.8468
Coral reef intertidal zone	Adaptive	0.9916	0.9949	0.9932
	DBSCAN	0.9643	0.9653	0.9648
	Statistical histogram	0.9928	0.9548	0.9734
	DRAGANN	0.9600	0.9841	0.9719
Estuarine intertidal zone	Adaptive	0.9953	0.9980	0.9967
	DBSCAN	0.9502	0.9191	0.9344
	Statistical histogram	0.9994	0.9308	0.9639
	DRAGANN	0.9907	0.9925	0.9916

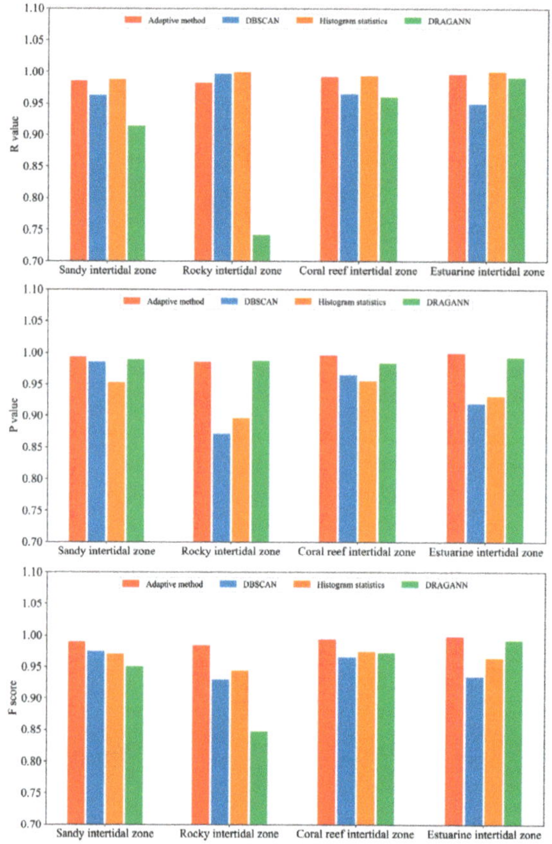

Figure 13. Quantitative statistical histogram of denoising performance.

Based on the statistical results, in the sandy intertidal zone, the statistical histogram algorithm achieved the highest recall rate of 0.9878, indicating that it can effectively retain signal photons. However, its precision was comparatively lower at 0.9522. In contrast, the adaptive denoising method achieved recall and precision rates of 0.9853 and 0.9930, respectively, with an F-score of 0.9891, the highest among all methods. This demonstrates the superior ability of the adaptive denoising method to balance recall and precision.

In the rocky intertidal zone, the statistical histogram algorithm once again achieved the highest recall rate of 0.9984, but its precision was relatively low at 0.8953. Similarly, the DBSCAN algorithm also had a high recall rate of 0.9968 but a low precision of 0.8716 and an F-score of 0.9300. The DRAGANN algorithm achieved the highest precision at 0.9870, but its recall rate and F-score were lower, at 0.7415 and 0.8468. The adaptive denoising method balanced recall and precision, achieving rates of 0.9818 and 0.9848, with an F-score of 0.9833, exhibiting strong adaptability and stability.

In the coral reef intertidal zone, the statistical histogram algorithm achieved a recall rate of 0.9928, a precision of 0.9548, and an F-score of 0.9734, indicating good denoising performance. The DRAGANN algorithm achieved a higher precision at 0.9841 but a lower recall rate of 0.9600 and an F-score of 0.9719. The adaptive denoising method achieved recall and precision rates of 0.9916 and 0.9949, with an F-score of 0.9932, demonstrating stronger adaptability and stability in the coral reef intertidal zone.

In the estuarine intertidal zone, the statistical histogram algorithm achieved a recall rate of 0.9994, almost retaining all signal photons, but a relatively low precision of 0.9308 and an F-score of 0.9630. The DRAGANN algorithm achieved a higher precision at 0.9925 while maintaining a high recall rate of 0.9907, resulting in an F-score of 0.9916, the best among all methods. The adaptive denoising method balanced recall and precision, achieving rates of 0.9953 and 0.9980, with an F-score of 0.9967, reflecting good denoising performance; it performed the best among all the methods. Table 4 shows the average R, P, and F of different methods across the four typical intertidal zones.

Table 4. The average metrics.

Denoising Method	*R*	*P*	*F*
Adaptive	0.9885	0.9927	0.9906
DBSCAN	0.9684	0.9353	0.9508
Statistical histogram	0.9946	0.9333	0.9628
DRAGANN	0.9016	0.9881	0.9401

By comprehensively comparing the denoising effects across various types of intertidal zones, it is evident that the traditional DBSCAN algorithm, the statistical histogram algorithm, and the DRAGANN algorithm can exhibit good denoising performance in specific environments, but it is difficult to maintain stability in different environments. In contrast, the adaptive denoising method demonstrates excellent performance and stability in various environments. It is able to achieve a good balance between recall and precision in most situations and maintains a high F-score. This indicates that the method can effectively adapt to the characteristics of different types of intertidal zones, achieving comprehensive and robust denoising.

5. Conclusions

This study addresses the challenges posed by the complex and diverse land cover types in intertidal zones and the difficulty in distinguishing between noise photons and signal photons. We propose a photon denoising method based on an adaptive strategy. The method's key feature is dynamically adjusting the denoising strategy by analyzing photon distribution characteristics at the water–land boundary in intertidal zones, employing corresponding denoising algorithms for different types of photon data. We selected ATL03 photon point cloud data from four typical intertidal zones for denoising analysis and compared the results with those of the traditional DBSCAN algorithm, the statistical histogram algorithm, and the DRAGANN algorithm. The following conclusions were drawn:

1. The proposed adaptive denoising method achieved excellent performance across the four typical intertidal zones, with average recall, precision, and F-score values reaching 0.9885, 0.9927, and 0.9906, respectively, demonstrating high overall accuracy. This indicates that the method can comprehensively remove noise while retaining the integrity of signal photons.
2. Compared to single traditional denoising algorithms, the adaptive denoising method exhibits significant advantages in handling complex terrain features. In areas with dramatic terrain changes, such as water–land boundaries, in particular, the method can more accurately identify and retain key topographic features while efficiently removing surrounding noise points.

Author Contributions: C.W. and S.L. proposed the methodology and wrote the manuscript. L.D. contributed to improving the methodology and is the corresponding author. L.C. helped edit and improve the manuscript. C.W. contributed to methodological testing. All authors have read and agreed to the published version of the manuscript.

Funding: This research was supported by the project of the Key Laboratory of SmartEarth, China (No. KF2023YB02-03), the project of the Natural Science Foundation of Hunan Province, China (No. 2023JJ30233), and the project supported by Scientific Research Fund of Hunan Provincial Education Department (Grant No. 24B0451).

Institutional Review Board Statement: Not applicable.

Informed Consent Statement: Not applicable.

Data Availability Statement: The ATL03 and ATL08 datasets provided in the article are available at https://search.earthdata.nasa.gov/, accessed on 20 April 2024.

Acknowledgments: We would like to express appreciation to the NASA NSIDC for providing the ICESat-2 data used in the article. Moreover, we thank the reviewers and editors for their helpful comments and insightful suggestions.

Conflicts of Interest: Author Lei Ding was employed by the company Space Star Technology Co., Ltd. The remaining authors declare that the research was conducted in the absence of any commercial.

References

1. Noujas, V.; Thomas, K.V.; Badarees, K.O. Shoreline management plan for a mudbank dominated coast. *Ocean Eng.* **2016**, *112*, 47–65. [CrossRef]
2. Xu, N.; Ma, Y.; Yang, J.; Wang, X.H.; Wang, Y.; Xu, R. Deriving Tidal Flat Topography Using ICESat-2 Laser Altimetry and Sentinel-2 Imagery. *Geophys. Res. Lett.* **2022**, *49*, e2021GL096813. [CrossRef]
3. Fitton, J.M.; Rennie, A.F.; Hansom, J.D.; Muir, F.M.E. Remotely sensed mapping of the intertidal zone: A Sentinel-2 and Google Earth Engine methodology. *Remote Sens. Appl. Soc. Environ.* **2021**, *22*, 100499. [CrossRef]
4. Kim, H.; Lee, S.B.; Min, K.S. Shoreline change analysis using airborne LiDAR bathymetry for coastal monitoring. *J. Coast. Res.* **2017**, *79*, 269–273. [CrossRef]
5. Renga, A.; Rufino, G.; D'Errico, M.; Moccia, A.; Boccia, V.; Graziano, M.D.; Aragno, C.; Zoffoli, S. SAR bathymetry in the Tyrrhenian Sea by COSMO-SkyMed data: A novel approach. *IEEE J. Sel. Top. Appl. Earth Obs. Remote Sens.* **2014**, *7*, 2834–2847. [CrossRef]
6. Li, J.; Zheng, K.; Li, Z.; Gao, L.; Jia, X. X-shaped interactive autoencoders with cross-modality mutual learning for unsupervised hyperspectral image super-resolution. *IEEE Trans. Geosci. Remote Sens.* **2023**, *61*, 5518317. [CrossRef]
7. Li, J.; Zheng, K.; Gao, L.; Ni, L.; Huang, M.; Chanussot, J. Model-informed multistage unsupervised network for hyperspectral image super-resolution. *IEEE Trans. Geosci. Remote Sens.* **2024**, *62*, 5516117. [CrossRef]
8. Zhong, J.; Sun, J.; Lai, Z. ICESat-2 and Multispectral Images Based Coral Reefs Geomorphic Zone Mapping Using a Deep Learning Approach. *IEEE J. Sel. Top. Appl. Earth Obs. Remote Sens.* **2024**, *17*, 6085–6098. [CrossRef]
9. Smith, B.; Fricker, H.A.; Holschuh, N.; Gardner, A.S.; Adusumilli, S.; Brunt, K.M.; Csatho, B.; Harbeck, K.; Huth, A.; Neumann, T.; et al. Land ice height-retrieval algorithm for NASA's ICESat-2 photon-counting laser altimeter. *Remote Sens. Environ.* **2019**, *233*, 111352. [CrossRef]
10. Neumann, T.A.; Martino, A.J.; Markus, T.; Bae, S.; Bock, M.R.; Brenner, A.C.; Brunt, K.M.; Cavanaugh, J.; Fernandes, S.T.; Hancock, D.W.; et al. The Ice, Cloud, and Land Elevation Satellite-2 mission: A global geolocated photon product derived from the Advanced Topographic Laser Altimeter System. *Remote Sens. Environ.* **2019**, *233*, 111325. [CrossRef]
11. Parrish, C.E.; Magruder, L.A.; Neuenschwander, A.L.; Forfinski-Sarkozi, N.; Alonzo, M.; Jasinski, M. Validation of ICESat-2 ATLAS Bathymetry and Analysis of ATLAS's Bathymetric Mapping Performance. *Remote Sens.* **2019**, *11*, 1634. [CrossRef]
12. Liu, J.; Liu, J.; Xie, H.; Ye, D.; Li, P. A Multi-Level Auto-Adaptive Noise-Filtering Algorithm for Land ICESat-2 Photon-Counting Data. *Remote Sens.* **2023**, *15*, 5176. [CrossRef]
13. Ester, M.; Kriegel, H.P.; Sander, J.; Xu, X. A density-based algorithm for discovering clusters in large spatial databases with noise. In Proceedings of the 2nd International Conference on Knowledge Discovery and Data Mining (KDD'96), Portland, OR, USA, 2–4 August 1996; pp. 226–231. Available online: https://dl.acm.org/doi/10.5555/3001460.3001507 (accessed on 22 May 2024).
14. Kui, M.; Xu, Y.; Wang, J.; Cheng, F. Research on the Adaptability of Typical Denoising Algorithms Based on ICESat-2 Data. *Remote Sens.* **2023**, *15*, 3884. [CrossRef]
15. Magruder, L.A.; Wharton, M.E.; Stout, K.D.; Neuenschwander, A.L. Noise filtering techniques for photon-counting ladar data. In Proceedings of the SPIE Defense, Security, and Sensing, Baltimore, MD, USA, 23–27 April 2012; p. 8379. [CrossRef]
16. Huang, X.; Cheng, F.; Wang, J.; Duan, P.; Wang, J. Forest Canopy Height Extraction Method Based on ICESat-2/ATLAS Data. *IEEE Trans. Geosci. Remote Sens.* **2023**, *61*, 1–14. [CrossRef]

17. Wang, L.; Zhang, X.; Zhang, Y.; Chen, F.; Dang, S.; Sun, T. A Density-Based Multilevel Terrain-Adaptive Noise Removal Method for ICESat-2 Photon-Counting Data. *Sensors* **2023**, *23*, 9742. [CrossRef] [PubMed]
18. Neuenschwander, A.; Pitts, K. The ATL08 land and vegetation product for the ICESat-2 Mission. *Remote Sens. Environ.* **2019**, *221*, 247–259. [CrossRef]
19. Leng, Z.; Zhang, J.; Ma, Y.; Zhang, J.; Zhu, H. A novel bathymetry signal photon extraction algorithm for photon-counting LiDAR based on adaptive elliptical neighborhood. *Int. J. Appl. Earth Obs. Geoinf.* **2022**, *115*, 103080. [CrossRef]
20. Xie, H.; Ye, D.; Xu, Q.; Sun, Y.; Huang, P.; Tong, X.; Guo, Y.; Liu, X.; Liu, S. A density-based adaptive ground and canopy detecting method for ICESat-2 photon-counting data. *IEEE Trans. Geosci. Remote Sens.* **2022**, *60*, 4411813. [CrossRef]
21. He, L.; Pang, Y.; Zhang, Z.; Liang, X.; Chen, B. ICESat-2 data classification and estimation of terrain height and canopy height. *Int. J. Appl. Earth Obs. Geoinf.* **2023**, *118*, 103233. [CrossRef]
22. Wang, K.; Li, H.; Zhang, N.; Zhang, J.; Zhang, X.; Gong, Z. Study on the Erosion and Deposition Changes of Tidal Flat in Jiangsu Province Using ICESat-2 and Sentinel-2 Data. *Remote Sens.* **2023**, *15*, 3598. [CrossRef]
23. Li, J.; Chu, S.; Hu, Q.; Cong, Y.; Cheng, J.; Chen, H.; Cheng, L.; Zhang, G.; Xing, S. Land-sea classification based on the fast feature detection model for ICESat-2 ATL03 datasets. *Int. J. Appl. Earth Obs. Geoinf.* **2024**, *130*, 103916. [CrossRef]
24. Zhong, J.; Liu, X.; Shen, X.; Jiang, L. A Robust Algorithm for Photon Denoising and Bathymetric Estimation Based on ICESat-2 Data. *Remote Sens.* **2023**, *15*, 2051. [CrossRef]
25. Wang, B.; Ma, Y.; Zhang, J.; Zhang, H.; Zhu, H.; Leng, Z.; Zhang, X.; Cui, A. A noise removal algorithm based on adaptive elevation difference thresholding for ICESat-2 photon-counting data. *Int. J. Appl. Earth Obs. Geoinf.* **2023**, *117*, 103207. [CrossRef]
26. Zhang, G.; Lian, W.; Li, S.; Cui, H.; Jing, M.; Chen, Z. A Self-Adaptive Denoising Algorithm Based on Genetic Algorithm for Photon-Counting Lidar Data. *IEEE Geosci. Remote Sens. Lett.* **2022**, *19*, 3067609. [CrossRef]
27. Lian, W.; Li, S.; Zhang, G.; Chen, X.; Li, Z. Denoising Algorithm Based on Local Distance Weighted Statistics for Photon Counting Lidar Point Data. In Proceedings of the IGARSS 2019–2019 IEEE International Geoscience and Remote Sensing Symposium, Yokohama, Japan, 28 July–2 August 2019; pp. 8988–8991. [CrossRef]
28. Zhang, J.; Kerekes, J.; Csatho, B.; Schenk, T.; Wheelwright, R. A clustering approach for detection of ground in micropulse photon-counting lidar altimeter data. In Proceedings of the Geoscience and Remote Sensing Symposium, Quebec City, QC, Canada, 13–18 July 2014; pp. 177–180. [CrossRef]
29. Zhu, X.; Nie, S.; Wang, C.; Xi, X.; Wang, J.; Li, D.; Zhou, H. A noise removal algorithm based on OPTICS for photon-counting LiDAR data. *IEEE Geosci. Remote Sens. Lett.* **2020**, *18*, 1471–1475. [CrossRef]
30. Xie, J.; Zhong, J.; Mo, F.; Liu, R.; Li, X.; Yang, X.; Zeng, J. Denoising and Accuracy Evaluation of ICESat-2/ATLAS Photon Data for Nearshore Waters Based on Improved Local Distance Statistics. *Remote Sens.* **2023**, *15*, 2828. [CrossRef]
31. Zheng, X.; Hou, C.; Huang, M.; Ma, D.; Li, M. A density and distance-based method for ICESat-2 photon-counting data denoising. *IEEE Geosci. Remote Sens. Lett.* **2023**, *20*, 6500405. [CrossRef]

MDPI AG
Grosspeteranlage 5
4052 Basel
Switzerland
Tel.: +41 61 683 77 34

Photonics Editorial Office
E-mail: photonics@mdpi.com
www.mdpi.com/journal/photonics

www.ingramcontent.com/pod-product-compliance
Lightning Source LLC
Chambersburg PA
CBHW061607210926
51850CB00009B/2243